Recent Advances in Biomechanics

Edited by Redha Taiar

Published in London, United Kingdom

IntechOpen

Supporting open minds since 2005

Recent Advances in Biomechanics
http://dx.doi.org/10.5772/intechopen.87371
Edited by Redha Taiar

Contributors
Paul Grimshaw, Christopher Jones, Merilyn Lock, Matthew Longbottom, Eliete Biasotto Hauser, Wyllians Vendramini Borelli, Jaderson Costa da Costa, Sarmad Shams, Muhammad Asif, Samreen Hussain, Paul-Dan Sirbu, Wilhelm Friedl, Dan Mihailescu, Liliana Savin, Andrei Scripcaru, Mihnea Theodor Sirbu, Mihaela Pertea, Razvan Cosmin Tudor, Norin Forna, Erik David García Lasluisa, Jefferson Michael Vela Rodríguez, Andrzej Jan Dyszkiewicz, Diana Hruby, Redha Taiar, Susan Z. Hua, Mohammad Mehdi Maneshi

First published in London, United Kingdom, 2020 by IntechOpen
IntechOpen is the global imprint of INTECHOPEN LIMITED, registered in England and Wales, registration number: 11086078, 7th floor, 10 Lower Thames Street, London, EC3R 6AF, United Kingdom
Printed in Croatia

British Library Cataloguing-in-Publication Data
A catalogue record for this book is available from the British Library

Additional hard and PDF copies can be obtained from orders@intechopen.com

Recent Advances in Biomechanics
Edited by Redha Taiar
p. cm.
Print ISBN 978-1-83968-071-7
Online ISBN 978-1-83968-072-4
eBook (PDF) ISBN 978-1-83968-073-1

For EU product safety concerns:
IN TECH d.o.o., Prolaz Marije Krucifikse Kozulić 3, 51000 Rijeka, Croatia, info@intechopen.com or visit our website at intechopen.com.

Meet the editor

Prof. Redha Taiar, Ph.D. Biomechanics is a Professor at the University of Reims, Champaign France. He is Head of the Redha Taiar Biomechanical Engineering Society developed for sport and medical advice for the industry. His research focuses on the industry of engineering for medicine and sport. He is an engineer for different industries such as Arena for sport and Sidas, Medicapteur for medical development. For industry workers, his last project was for the Notrax Society (Conception and validation of anti-fatigue mats). For sport, his last project focused on the development of a swimsuit for triathlons and swimming for Brazil Olympic Games (2016) and the suit fabrics for skiing at the Olympic Games at Sotchi in 2014. He is a specialist in the biomechanics of health disease and rehabilitation.

Contents

Preface

Writing or managing a scientific book, as it is known today, depends on a series of major factors such as selecting researchers, reviewing chapters, communicating with the researchers, and motivating the authors to achieve this objective of this publication. The idea of this book came after many years of work in biomechanics, health disease, and rehabilitation. From the exchange with the authors (four groups of authors from four different countries), we learned much from each other and we decided with the publisher to transfer this knowledge to readers interested in the impact of biomechanics today on the analysis of movement and optimization. The main objective is to include some chapters on the scope of biomechanical analysis and technologies in human behavior tasks. Biomechanics is present in our lives in different situations. The combination of the principles of mathematics, functional anatomy, and mechanics allows biomechanics to explore and understand biological problems. The tools and methods are able to quantify and improve the discriminate parameters characterizing movement in different cases such as at the sport level, work, and patients' daily lives. The concern is to effectively combine and coordinate biomechanical research and results in order to understand and improve human mechanics in different situations in our daily lives. For instance, with biomechanics and in correlation with other approaches we can analyse the contribution of tissues and organs that are responsible for various important functions, such as heart rate, peristaltic movement of the digestive tract, contractions of the blood vessels and skeletal muscle contractions. In addition, during the various daily activities, such as walking or running, biomechanics quantify the impact of the external environment that can transmit energy to the organs and tissues of the human body permitting efficiency. This book presents an understanding of biomechanics through chapters analyzing human behavior in sport and from a medical perspective. It offers a comprehensive range of principles, methods, techniques, and tools to provide the reader with clear knowledge of the impact of biomechanic processes. The text considers physical, mechanical, and biomechanical aspects and is illustrated by different key application domains such as sports performance, sports science, ergonomy science, gait and human posture, and musculoskeletal disorders in medicine.

Apart from the introduction, the first three chapters provide useful tools for measuring, generating, simulating, and processing in biomechanics with the clinical and experimental applications in medicine. The last section describes the application of biomechanics in sports performance. Engineers, researchers, and students from biomedical engineering and health sciences, as well as industrial professionals, can profit from this compendium of knowledge about biomechanics applied to the human body.

Acknowledgements

We express our thanks to all authors for the successful collaboration in the interest of this book and we give you recognition for everything you have done to achieve this objective. Thank you colleagues for your kind attention.

Redha Taiar
Université de Reims Champagne-Ardenne,
Reims, France

Section 1

Introduction

Chapter 1

Introductory Chapter: Biomechanics, Concepts and Knowledge

Redha Taiar

1. The concepts and basis of biomechanics

The achievement of the human voluntary movement is seemingly simple, but rather it is considerably complex. As it is a very complex mechanism which allows many nerve structures to make decisional and/or reflexional choices, then this mechanism "defines" and "controls" the movement, through the nerve impulses intended for the musculoskeletal system. In the human body, around 640 muscles are involved with 213 bones rigidifying several segments ranging from a few cm^3 to several thousand cm^3 through more than 100 joints. The whole system thus allows greater than 100 degrees of freedom. Biomechanics study the complexity of the human body through behavior and performance in daily life with respect to musculoskeletal system with the aim to optimize the system [1, 2]. This discipline seems essential in today's world and considerably promising for the future. Indeed, the concepts, methods, and analytical techniques that we use to characterize human mechanics represent major economic stakes. The development is necessarily inscribed in a vision of interactions between the physical sciences (metrology, complex mechanical and electronic systems), information sciences and technologies, and life sciences (materials, tissues, organs, and limbs) [3–8]. The ambition of the research work carried out in biomechanics is to improve the performance of high-level athletes and the comfort and quality of life of patients and to minimize stress on joints in real-field or laboratory situations. Biomechanics is subdivided into kinematic (**Figure 1**) and kinetic (**Figure 2**) analyses. Kinematics is concerned with the analysis of motion (e.g., to determine the forces applied to a joint from the inverse dynamics), while kinetics studies the forces that cause or result from it (e.g., the reaction of the ground when walking).

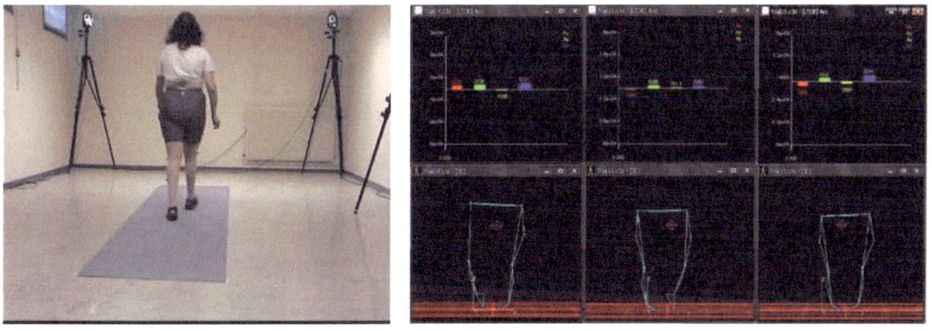

Figure 1.
Kinematic device for 3D motion analysis. The example of a patient with myopathy.

IntechOpen

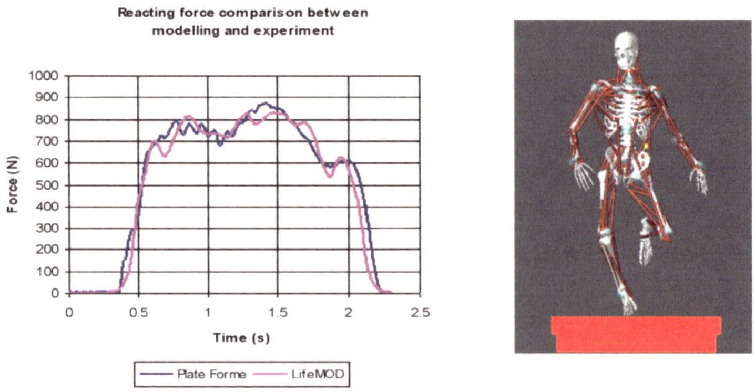

Figure 2.
Kinetic analysis of motion. Correlation between modeled and experimental data.

2. Modelization in biomechanics

Different models can be considered ranging from the human body represented by its center of gravity to the model integrating both motor control and musculoskeletal modeling of the human body. With the current medical techniques (Scanner, MRI, and X-ray) and recent computer modeling, many technical and scientific advances are now possible in biomechanics [9]. The aim is to modelize mathematically (**Figure 3**) and simulate the mechanical behavior of the human body under the application of various constraints. This model will be correlated with cases of declared pathologies by considering behavioural control as a main objective of prevention (**Figure 4**). The simulation will make it possible to predict the appearance of pathologies that may slow down the stability or progression of human mechanics in all combined fields [10–12]. The recommendations will be applicable with the aim to optimize human mechanics.

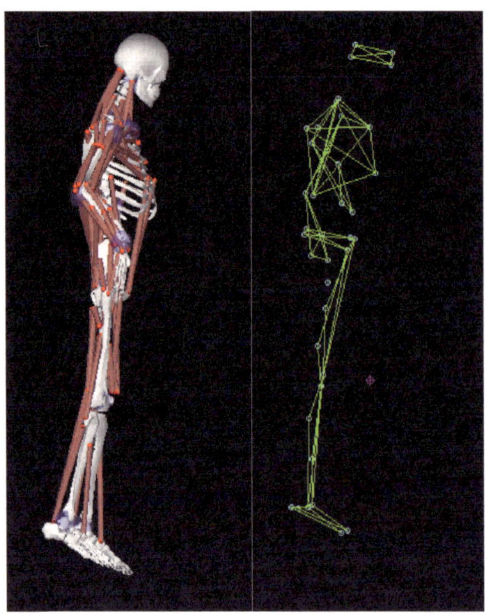

Figure 3.
Dynamic modeling from kinematic data, example of a vertical jump.

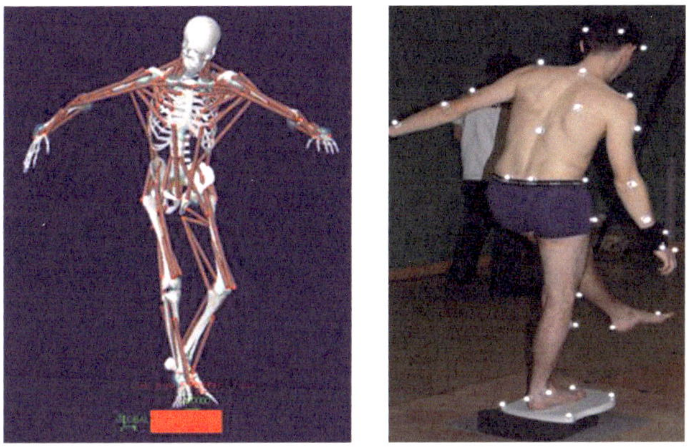

Figure 4.
Dynamical simulation permitting the optimization of the movement.

Mathematical modeling in life sciences or medical sciences is hardly developed. This modeling involves applying physical laws to analyze both human and animal movements and to quantify and analyze the discriminating parameters of movement. Given its very complex approach, "skeletal" modeling consists of representing the body by a certain number of segments (often considered indeformable to simplify calculations). The interest of this modelling lies in the possibility of combining and coordinating research results [9] with an efficient way in innovative projects oriented towards CAD—simulation—rapid prototyping (**Figure 5**). Applications will be in medicine (e.g., development of new orthotics) and in sports (e.g., proposing a methodology for optimizing sports clothing).

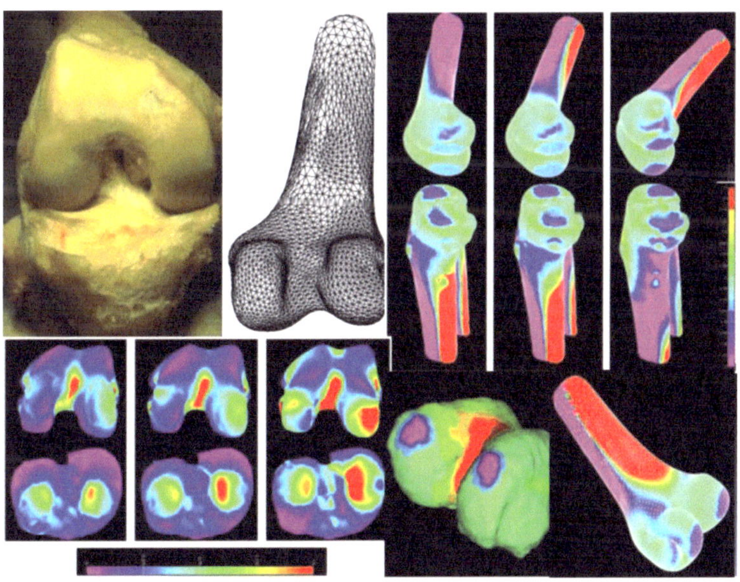

Figure 5.
Anatomy (image formation), CAD, mesh, and finite element analysis of the knee joint. Procedure and quantification of mechanical stress at the joint level.

Author details

Redha Taiar
Université de Reims Champagne-Ardenne, Reims, France

*Address all correspondence to: redha.taiar@univ-reims.fr

References

[1] Fourchet F, Kelly L, Horobeanu C, Loepelt H, Taiar R, Millet G. High-intensity running and plantar-flexor fatigability and plantar-pressure distribution in adolescent runners. Journal of Athletic Training. 2015;**50**(2):117-125

[2] Fourchet F, Horobeanu C, Loepelt H, Taiar R, Millet G. Foot, ankle, and lower leg injuries in young male track and field athletes. International Journal of Athletic Therapy and Training. 2011;**16**(3):19-23

[3] Popa CV, Zaidi H, Arfaoui A, Polidori G, Taiar R, Fohanno S. Analysis of wall shear stress around a competitive swimmer using 3D Navier-stokes equations in CFD. Acta of Bioengineering and Biomechanics. 2011;**13**(1):3-11

[4] Popa CV, Arfaoui A, Fohanno S, Taïar R, Polidori G. Influence of a postural change of the swimmer's head in hydrodynamic performances using 3D CFD. Computer Methods in Biomechanics and Biomedical Engineering. 2014;**17**(4):344-351

[5] Polidori G, Renard Y, Lorimier S, Pron H, Derruau S, Taiar R. Medical infrared thermography assistance in the surgical treatment of axillary Hidradenitis Suppurativa: A case report. International Journal of Surgery Case Reports. 2017;**34**:56-59

[6] Alexandre D, Prieto M, Beaumont F, Taiar R, Polidori G. Wearing lead aprons in surgical operating rooms: Ergonomic injuries evidenced by infrared thermography. Journal of Surgical Research. 2017;**209**:227-233

[7] Polidori G, Marreiro A, Pron H, Lestriez P, Boyer FC, Quinart H, et al. Theoretical modeling of time-dependent skin temperature and heat losses during whole-body cryotherapy: A pilot study. Medical Hypotheses. 2016;**96**:11-15

[8] Beaumont F, Taiar R, Polidori G, Trenchard H, Grappe F. Aerodynamic study of time-trial helmets in cycling racing using CFD analysis. Journal of Biomechanics. 2018;**67**(23):1-8

[9] Taïar R, Fogarassy P, Boyer F, Lodini A. Knee joint distribution: 3D finite element analysis. Series on Biomechanics. 2010;**25**(3-4):3-11 ISSN: 1313-2458

[10] Sá-Caputo DC, Paineiras-Domingos LL, Oliveira R, Neves MFT, Brandão A, Marin PJ, et al. Acute effects of whole-body vibration on the pain level, flexibility, and cardiovascular responses in individuals with metabolic syndrome. Dose-Response. 2018;**16**(4)

[11] Sá-CaputoD,Paineiras-DomingosLL, Francisca-Santos A, Dos Anjos EM, Reis AS, Neves MFT, et al. Whole-body vibration improves the functional parameters of individuals with metabolic syndrome: An exploratory study. BMC Endocrine Disorders. 2019;**19**(1):6

[12] Dany A, Barbe C, Rapin A, Réveillère C, Hardouin J-B, Morrone I, et al. Construction of a quality of life questionnaire for slowly progressive neuromuscular disease. Quality of Life Research. 2015;**24**(1):2615-2623

Section 2

Fundamental Biomechanics

Chapter 2

Muscle Mechanics and Electromyography

Sarmad Shams, Muhammad Asif and Samreen Hussain

Abstract

This chapter will begin with the introduction to biomechanics and its relationship with the physiology and anatomy. Then it introduces the basic concepts of kinematics, kinetics, and anthropometry and discusses in detail the muscle mechanics and electromyography. The muscle is the actuator of the human body, especially the skeletal muscles which are attached with the skeleton play an important role in defining the movements of the human body. The human body controls the muscle through the nervous system, and this nervous system generates signals called electroencephalogram (EEG) which upon leaving the nerves excites the muscle and converted into muscle signals usually called electromyogram (EMG). In this chapter, we will discuss the mechanics of the muscle in conjunction with the EMG. EMG is the tool to study the activity of the muscles and hence the key to understand the mechanics of the human body.

Keywords: electromyogram (EMG), biomechanics, muscle mechanics

1. Introduction

What is Biomechanics? The biomechanics is the study of the structure and function of biological systems. "Bio" means "living" and "Mechanics" is "Forces and its

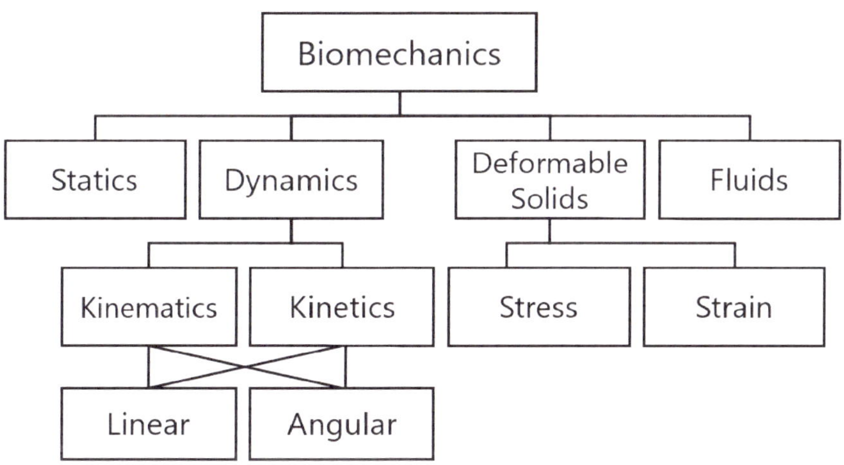

Figure 1.
The division of mechanics in a nutshell.

effects," hence "Biomechanics," terms evolved which literally means the forces and its effects on the living cells. The biomechanics includes the study of tissues including bones, cartilages, ligaments, tendons, muscles, and nerves, studying all these tissues at different levels and scales ranging from the single cell to tissues, to organs, and to the whole body level.

The biomechanics is divide into following **Figure 1**.

1.1 Relationship with physiology and anatomy

To study the biomechanics, one should know the basic of human structure and its functions, that is, anatomy and physiology of the human body.

2. Essentials of biomechanics

2.1 Kinematics

Kinematics is the study of motion without involving the factor that causes that motion. The whole kinematics theory and analysis will evolve around three basic variables of mechanics, which are as follows:

1. Displacement
2. Velocity
3. Acceleration

2.2 Kinetics

Kinetics is the study of motion, and the elements causes that motion, that is, force and torque. The kinetics as evident from the name kinetics will discuss the mechanics of the human body that are under motion, and the three basic variables that cause or influence the motion are as follows:

1. Force
2. Torque
3. Mass

2.3 Anthropometry

Anthropometery is a combination of two Greek words anthropos, which means "Human" and metron, which means "Measure." Therefore, the literal meaning of the anthropometry is the measurement of the human individual. When discussing the mechanics of the human body, the discussion and understanding of the anthropometry become essential, due to the fact that the mechanics cannot be complete without the measurement of the subject. The basic parameters under consideration when discussing the anthropometry are *Height, Weight, Size* and *Esthetics* of organs, limbs, and other body parts. A number of studies were carried out for the collection and compiling a database to standardize the anthropometry data, such as [1–3].

3. Muscle mechanics

3.1 Muscular system

Contrary to the bones, which provide the structural strength and stability to the human body, muscles are responsible to hold the bones with the possible movement of each joint with the application of forces by contracting themselves. Muscles are not simply the force generators in the body; the force developed in the muscle depends on the level of neural excitation at the central nervous system (CNS) with the speed and length at which the muscle contracts. Almost all the muscles are arranged in pairs as agonistic and antagonistic muscles, and these pairs work against each other, as one muscle pull then the other muscle relaxes and stretched in response. There are almost 320 muscle pairs that exist in a human body which make up almost 40% of the whole body mass.

4. Electromyography

Electromyography is the study of electromyograph (EMG). EMG is the recording of the electrical activity of the muscle. This electrical activity is produced due to the ionic movement in the muscle in response of the conducting signal from the motor unit of a neuron. This motor unit (MU) acquires single or multiple muscle fibers to perform the actuation of the muscle; this selection of the muscle fiber is random and it is not necessary that the previously acquired MU will acquire the same muscle fiber.

5. EMG signal

Electromyogram (EMG) is the electrical activity produced by a contracting muscle. EMG signal is extensively used in the field of rehabilitation, biomechanics, orthopedics, ergonomic product design, and prostheses. Due to the fact that EMG allows directly looking into the muscle and measuring the muscular performance, it also helps in decision-making both before and after surgery. The basic functional element that is responsible for producing the EMG signal is called the motor unit (MU). The MU consists of an $\alpha - motorneuron$ that has cell body in the spinal cord and extends its axon from the spinal cord to the skeletal muscle fiber (as shown in the **Figure 2**), where it innervates and forms a junction, usually called motor end-plates.

The signal from the $\alpha -$ motorneuron causes depolarization in the muscle fiber that travels in either direction from the junction and creates a potential difference. This difference is measured by the electrodes. The muscle fibers connected to a single neuron react together and hence termed as motor unit. The signal generates by the MU is called motor unit action potential (MUP). The resulting EMG signal is the sum of all activated MUs during the contraction of that muscle. When the signal is acquired using surface electrodes, the signal has to travel through the remaining tissues before reaching the electrodes. This traveling of the signal results in the decaying of the signal amplitude.

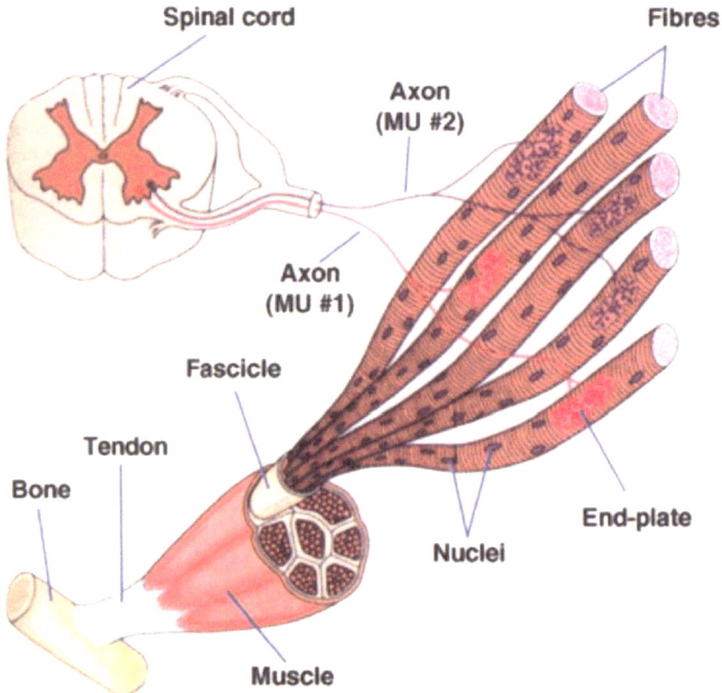

Figure 2.
A typical α – motorneuron extends from the spinal cords and ends at the motor end-plates in the skeleton muscle fiber [4].

6. Acquisition of the EMG signal

The signal generates due to the contraction of the muscle can be detected by placing the electrode on the surface of the muscle. The electrode placement is the main challenge in the acquisition of the surface EMG (sEMG), as the strength of EMG signal varies significantly if the electrodes are slightly displaced from the previous position [5]. In this research, the electrodes have been placed on the triceps and biceps muscles. The placement on the triceps muscle is quite a challenge due to its small size. We searched the spot on the subject for the best EMG signal acquisition. The signal detects by the surface electrodes can be seen in the **Figure 3a** and termed as a raw EMG signal.

The raw EMG signal is then passed through an instrumentation amplifier with high common mode rejection ratio (CMRR). The instrumentation amplifier is configured in differential amplification mode to eliminate the noises using the CMRR feature. The gain of the instrumentation amplifier is set as high as 1000 to amplify the minute EMG signal, as the typical EMG signal amplitude measures around 100 μV. After successfully eliminating the common noises and amplification, the amplified EMG signal is passed through a bandpass filter of a low cutoff frequency of 450 Hz and a high cutoff filter of 10 Hz for further filtration. This filtered EMG signal is then passed through a notch filter to eliminate the 50 Hz line frequency. The final filtered EMG signal is then rectified and passes through a smoothing filter to obtain the processed EMG signal. This signal is then used by the controller to execute the control algorithm. **Figure 4** shows the flow of the EMG signal acquisition and process while **Figure 5** shows the EMG signal obtained at each step of the EMG signal acquisition, that is, after the acquisition, then filtration, then rectification, and then smoothing process (moving average filter and RMS).

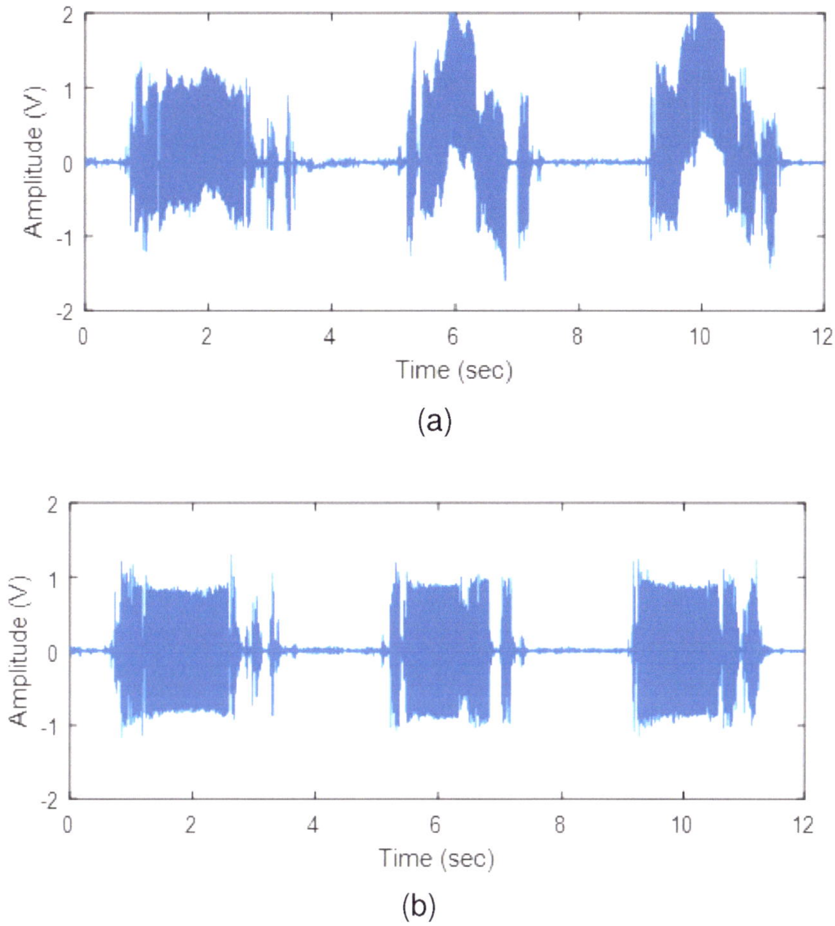

Figure 3.
The EMG signal at every step of the EMG signal acquisition. (a) Raw EMG of three biceps contraction; (b) EMG signal after filtration process.

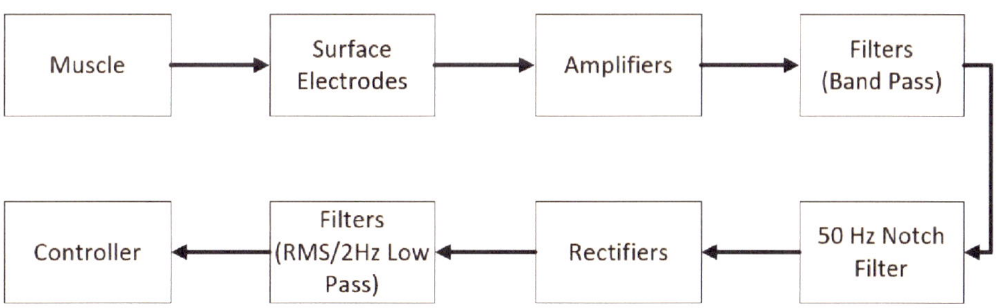

Figure 4.
sEMG signal acquisition flow chart.

6.1 Electrode placement for EMG signal

Since 1980, researchers are working on the development of protocols and standards for the surface electromyogram (sEMG) electrode placement procedures. Initial attempt was done by the International Society of Electrophysiological Kinesiology in 1980 [6]. After few years, more detailed and in-depth report was published as surface EMG for a non-invasive assessment of muscles (SENIAM).

This report was published with the support of European Union. SENIAM was reviewed and further deliberated by a number of laboratories around the world and more refined booklet was published as SENIAM 8: European Recommendations for Surface Electromyography, 1999 [7]. Few of SENIAM recommendations are as follows:

- sEMG sensor—In sEMG sensor, the SENIAM recommend the electrode shape, size, and constructions for different muscle size and volume. This category also includes details of electrode material and electrode inner distance, etc.

- Sensor placement—The sensor placement is the most critical part of the sEMG signal acquisition during the preparation of the subjects' skin and selection of sensor location. This category also includes the standards for the placement and fixation and testing of the connections.

- Sensor locations—The last and final standardized category in the SENIAM is the optimal sensor location for the best sEMG signal with minimal crosstalk and artifacts. Almost all the upper and lower body details can be found in this catalog.

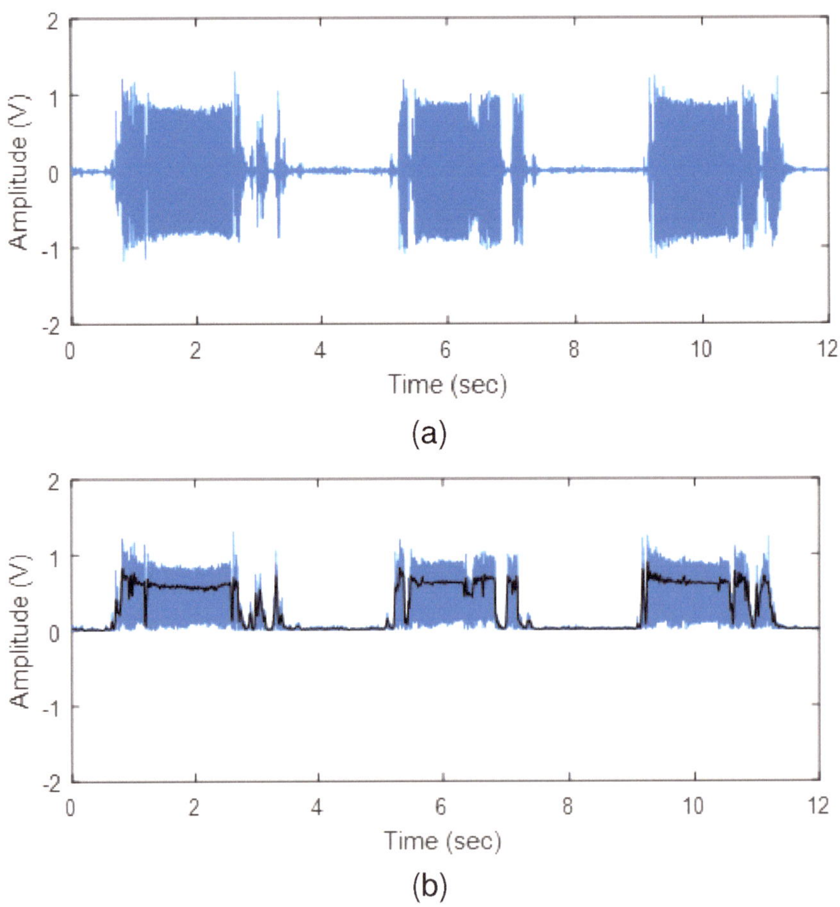

Figure 5.
The EMG signal (a) after band pass filter (b) after rectification in blue and EMG signal after smoothing filter in black.

6.2 Selection of amplifier gain

Amplifier gain plays an important role in the acquisition of quality and noise-free sEMG signal after the selection of electrode and its placement. The sEMG signal of a normal adult ranges from few hundreds microvolts to 2 mV. Contrary, in athletes, the sEMG signals 5 mV is recorded during maximum voluntary muscle (MVC) [8]. Selection of the amplifier gain considerably depends on the application and requirement of the system. However, for 2 mV input sEMG signal, a gain of 1000 can be used. This gain will results in an output of 2 V. This amount of amplification will also lead in the magnification of noises which can be ignored otherwise. To overcome the amplification of noises acquired with the sEMG signal, one can use the electrodes in differential configuration. In a typical differential amplifier, two electrodes are used with one common or ground electrode. The two electrodes are connected with the positive and negative terminal of an amplifier as shown in **Figure 6**. A typical differential configuration of operational amplifier (op-amp) is shown in **Figure 6**. The gain of the op-amp in differential configuration can be set using the Eq. (1). Whereas the output voltage V_{out} can be calculated with reference to the input voltages V_1 and V_2 using Eq. (2) if both the input and feedback resistors as shown in **Figure 6** are same.

$$A_v = \frac{V_{in}}{V_{out}} = \frac{V_{out}}{V_2 - V_1} \tag{1}$$

$$V_{out} = \frac{R_f}{R_i}(V_2 - V_1) \tag{2}$$

The main advantage of using differential configuration op-amps is to eliminate the noises common to both the input pins. The common noises will be canceled out due to the positive and negative input terminals resulting in a phenomenon known as common mode rejection ratio (CMRR) which is discussed in detail in the Section 6.4.

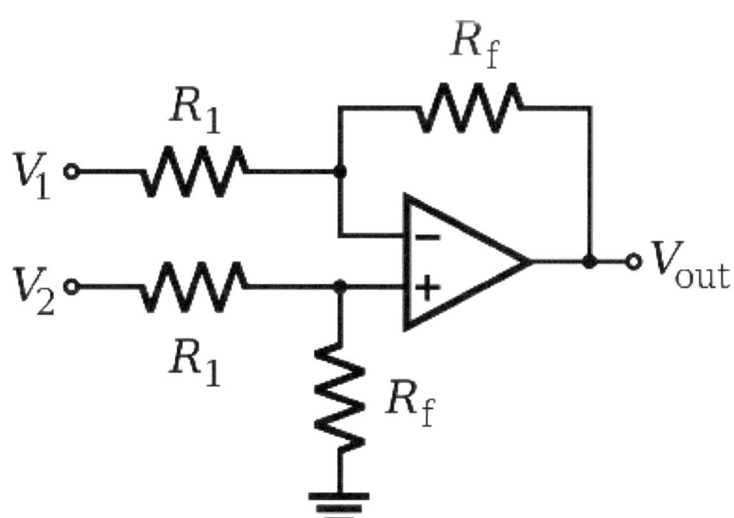

Figure 6.
A typical operational amplifier in differential amplifier configuration.

6.3 Input impedance

Human skin behaves different to different things (such as gel, oil, cream, drugs), especially current and voltages of different frequencies and amplitudes [9]. The resistance experienced by the electrodes during the acquisition of the sEMG signal can be reduced by using the gel in between the electrodes and skin [10].

6.4 Common mode rejection of noises

The amplification circuit has an ability to reject the common signal to both of its input if the signal is in phase and has same amplitude. This property of the amplifier is referred to as common mode rejection ratio (CMRR). In the acquisition of the biopotentials, like EEG, ECG, and sEMG, this property of the amplifier comes in handy if it is used in the differential configuration. Almost, all the op-amps inherit this property with the variation of the CMRR value. Ideally, an op-amp should have infinite amount of CMRR, but in real, it is limited to the range of 70–120 dB [11]. The typical formula to calculate CMRR is shown in Eq. (3). Where A_{DM} is the *differential mode gain* and A_{CM} is the *common mode gain*. Eq. (4) shows the CMRR formula to convert it into *deci Bell*.

$$CMRR = \frac{A_{DM}}{A_{CM}} \tag{3}$$

$$CMR(dB) = 20\log_{10}(CMRR) \tag{4}$$

6.5 Crosstalk of the EMG signal

It is quite difficult to identify the muscle when the sEMG signal acquired from the skin surface which contains a bunch of muscles underneath especially the forearm muscles. A cross-sectional view of the forearm is shown in **Figure 7**. The sEMG signal acquired from the forearm will be resulting from the contraction of multiple muscles due to the reason that the muscles are bundled and overlap each other in the forearm area. It is quite impossible to avoid crosstalk in this area as the active small motor unit range is around 0.5 cm and large motor unit range is around 1.5 cm [12]. Therefore, the electrode placed on the surface will acquire the sEMG signal produced by the contraction of multiple muscles' motor units which are under it. This crosstalk can be minimized by the following techniques:

- Manually checking the muscle resistance
- Using cross-correlation technique of signal processing.

Manual identification of the muscle resistance is not feasible in all cases. Therefore, the most successful technique is to use the cross-correlation method [13]. A general formula for the detection of cross-correlation is shown in Eq. (5); however, detail discussion is beyond the scope of the book.

$$R_{xy}(\tau) = \frac{\frac{1}{T}\int_{0}^{T} x(t)\,y(t-\tau)\,dt}{\sqrt{R_{xx}(0)R_{yy}(0)}} \tag{5}$$

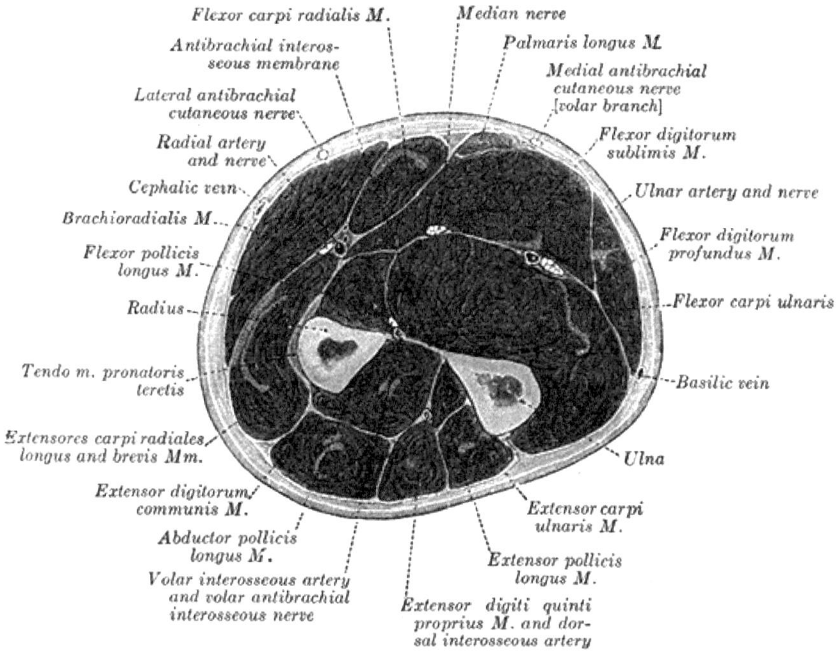

Figure 7.
Cross-sectional view of human forearm, image courtesy [14].

7. EMG signal processing

After the acquisition and amplification of the sEMG signal, the signal processing has been started as shown in **Figure 4**. The following blocks are essential for the signal processing of the sEMG signal, and the sequential process is shown in **Figure 8**.

- Filters (bandpass)
- Notch Filter (50 Hz)
- Rectification
- Smoothing filters

7.1 Filters—bandpass

Not only the amplitude of the sEMG signal depends on the site of acquisition but also the frequencies vary with it. Other factors depend on the subject physical health as the sEMG signal of athletes has higher amplitude and frequencies as compared with the normal subject [8]. The recommended cutoff frequencies setting for the bandpass filter are from 10 to 500 Hz [7]. However, in some cases, a tight band of 10–250 Hz may be used.

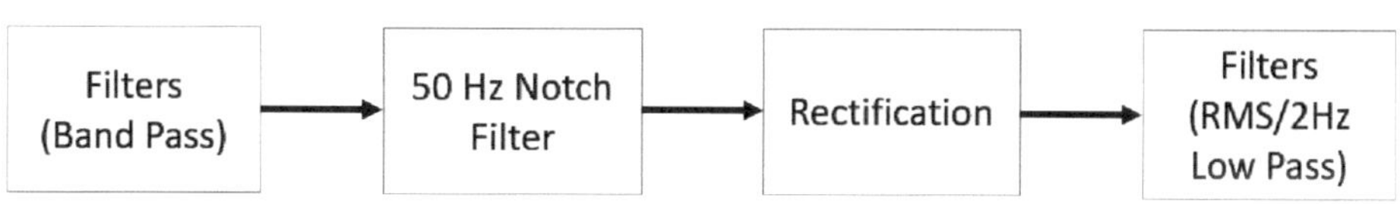

Figure 8.
The signal processing blocks in a typical sEMG acquisition system.

7.2 Rectification

In electrical, rectification is referred to a process that converts the alternating current (AC) to direct current (DC). In signal processing, the rectification is referred to convert the negative values of a signal into positive. In other words, the rectification is similar to take the absolute value of the signal. **Figure 5a** shows the signal of sEMG obtained during the contraction of bicep muscle. The rectified sEMG signal can be seen in **Figure 5b**. The signal in the black color (in **Figure 5b**) is obtained after the smoothing filter which will be discussed in the next section.

7.3 Smoothing filters

The sEMG signal is random and cannot be reproduced again with same amplitude, frequency, and shape. Therefore, sEMG undergoes signal smoothing techniques to minimize the effect of the non-reproducible part of the signal. Mostly, following techniques are used for smoothing the sEMG signal and one can select anyone of the techniques depending on the system requirements.

7.3.1 Root mean square (RMS)

The root mean square or RMS is the most commonly used technique among the signal processing community for the smoothing of signal. The RMS calculation is based on the square root and the mean power of the signal as shown in the Eq. (6).

$$RMS = \sqrt{\frac{\sum_{i=1}^{n} x_i^2}{N}} \tag{6}$$

7.3.2 Moving average filter (MA)

As the name depicts, the moving average (MA) filter takes average of the samples as it moves forward. In other words, a window is set defined for a specific number of samples, then the data in the window are averaged by the sliding window technique resulting in the smoothing of the signal. The MA filter can be applied on the sEMG signal using Eq. (7).

$$y[i] = \frac{1}{M} \sum_{j=0}^{M-1} x[i+j] \tag{7}$$

7.3.3 Low pass filter

Another common technique for smoothing of the sEMG signal is to use a low pass filter that blocks all the higher frequencies of the signal. For this purpose, usually, a low pass filter with a cutoff frequency between 2 and 5 Hz is used.

8. Conclusions

The biomechanics is basically divided into four major areas: kinetics, kinematics, anthropometery, and electromyography. To understand the biomechanics, one must have clear understanding of all the four major areas especially the EMG. Since, the main application of the biomechanics is to observe the problem and improve the diagnosis and fixation of any disease or lacking identified during the analysis. Researchers have developed a number of techniques to monitor the behavior of human mechanics using 3D cameras, motion sensors, reflector markers but none of the technique gave the essential analysis which can be obtained through EMG. Therefore, this chapter covers the essential knowledge required to build a knowledge base for the reader for the better understanding of the Biomechanics of human body.

Author details

Sarmad Shams[1*†], Muhammad Asif[2†] and Samreen Hussain[3]

1 Sir Syed University of Engineering and Technology, Karachi, Pakistan

2 Ziauddin University, Karachi, Pakistan

3 Begum Nusrat Bhutto Women University, Sukkur, Pakistan

*Address all correspondence to: sarmadshams@yahoo.com

† These authors contributed equally.

References

[1] Hall C, Quigley O, Giles R, Sc M. Upper limb anthropometry: The value of measurement variance studies. The American Journal of Clinical Nutrition. 1980;**33**(8):1846-1851

[2] McDowell MA, Fryar CD, Ogden CL, Flegal KM. Anthropometric reference data for children and adults: United States, 2003-2006. National Health Statistics Reports. 2008;**10**:2003-2006

[3] Greiner TM. Hand Anthropometry of U.S. Army Personell. Technical Report Natick. 1991;TR-92/011:434. Available from: http://oai.dtic.mil/oai/oai?verb=getRecord{&}metadataPrefix=html{&}identifier=ADA244533

[4] Muzumdar A. Powered Upper Limb Prostheses: Control, Implementation and Clinical Application; Verlag Berlin Heidelberg: Springer; 2004

[5] Hermens HJ, Freriks B, Disselhorst-Klug C, Rau G. Development of recommendations for SEMG sensors and sensor placement procedures. Journal of Electromyography and Kinesiology. 2000;**10**(5):361-374. Available from: http://linkinghub.elsevier.com/retrieve/pii/S105064110000027

[6] Winter DA, Rau G, Kadefors R, Broman H, De Luca C. Units, Terms and Standards in the Rporting of EMG Research. Report by the Ad hoc Committee of the International Society of Electrophysiological Kinesiology; 1980

[7] Hermens HJ, Freriks B, Merletti R, Stegeman D, Blok J, Rau G, et al. European Recommendations for Surface ElectroMyoGraphy. Netherland: Roessingh Research and Development; 1999. pp. 8-11

[8] #Book_Lib. The ABC of EMG. March; 2006. Available from: http://www.noraxon.com/docs/education/abc-of-emg.pdf

[9] Spach MS, Barr RC, Havstad JW, Long EC. Skin-electrode impedance and its effect on recording cardiac potentials. Circulation. 1966;**34**(4):649-656

[10] Tronstad C, Johnsen GK, Grimnes S, Martinsen ØG. A study on electrode gels for skin conductance measurements. Physiological Measurement. 2010;**31**(10):1395-1410

[11] Pallás-Areny R, Webster JG. Common mode rejection ratio in differential amplifiers. IEEE Transactions on Instrumentation and Measurement. 1991;**40**(4):669-676

[12] Fuglevand AJ, Winter DA, Patla AE, Stashuk D. Detection of motor unit action potentials with surface electrodes: Influence of electrode size and spacing. Biological Cybernetics. 1992;**67**(2):143-153

[13] Winter D, Fuglevand AJ, Archer S. Crosstalk in surface electromyography: Theoretical and practical estimates. Journal of Electromyography and Kinesiology. 1994;**4**(1):15-26

[14] Chaudhry M, Arain A. Anatomy, Shoulder and Upper Limb, Forearm Compartments. 2019

Chapter 3

Theoretical Biomechanics: Design of the Associated Measurement Symmetry System

Andrzej Jan Dyszkiewicz and Diana Hruby

Abstract

Based on the own experience of the authors and the literature, an original proposition of the device and system software based on the quotient scale was presented, where the mutual relations of parameters from several or several simultaneously working measuring devices are easily visible on a common graph in real time. The project is connected with an outline of problems of musculoskeletal system diagnostics, aiming at creating a universal, systemic parametric graph, defining in parallel-multi-parametric and quantitative manner, the patient's initial health profile as a set of parametric symmetries (or asymmetries), having specific and separate characteristics for healthy individuals in all age groups, as well as in groups with specific disease units. The directional pattern for Authors, aiming at the systemic recognition of the causative phenomena and the consequences of motion in the form of a multi-parameter graph calibrated by a common time axis is the way of electrode location, recording and parameterization of the ECG curve. This universal formula, defining parametrically the problems of evolutionary and involutionary norms, as well as most known pathologies, has become the foundation of cardiology, functioning worldwide over the language barrier.

Keywords: AMSS, parametric symmetries, quotient scale, systemic multiparametric graph

1. Design of the associated measurement symmetry system

The questions posed above express doubts and searches of many specialists in the field of biomechanics - anthropomotorics in particular. Initiation of the associated measurement symmetry system (AMSS) was an attempt not only to answer the above questions but also to draw attention of the medical and scientific community involved in the study of the musculoskeletal system, to the need to search for a universal standard (protocol) and a multiband system for evaluation of the structural and motor characteristics of the body. The analysis of the existing measuring instruments clearly indicates that in order to simultaneously use several measurement methods, each of which has originally different system of units (on an interval scale), we are not able to assess the interparametric relationships in real time, and possible relationships between parameter trends can be expressed with a long delay, by means of using statistical tests only. In this situation, a good and expected solution would be to illustrate on one graph, calibrated with one time stamp (in a selected, universal frame of reference), a whole range of parameters, starting with the traditional vector features of motion, with attached, parallelly registered causative parameters of motion, e.g., rheological, impedance, and magnetic, as well as parameters of physical effects of motion on the ground (strain gauge measurements) and electrodiagnostic imaging and psychometrics. Reflecting on one correlated graph, the phenomena of parameter symmetry for parallelly used measurement techniques (or with a slight time shift) provide a unique opportunity to formulate static and dynamic motor standard of the body, containing the parameters of the specific psychometric and electrogalvanic predispositions, resulting in the creation of a time-space distribution of motion vectors and also the temporal-spatial distribution of gravity vectors in the area of contact of the limbs with the ground [1–4].

The use of the quotient scale allows to incorporate into the system almost every research method, in a manner that refers to the symmetrical construction plan of the musculoskeletal system (and many internal organs) or, in other words, to measure the mirror areas of the left and right upper and lower limbs and trunk. The consequence of this action is determining the nondimensional representation (coefficient) for the symmetry of a measurement pair in the field of a specified physical quantity, presented as a symmetry, asymmetry, or lateralization indicator, whose value is in the range of 0–1.

The next part of the work presents the author's design of a device implementation, along with the base of a systemic software based on the quotient scale, where interactions between parameters derived from several concurrent measurement devices are easily visible on one graph in real time. It should be noted that the basic feature of the quotient scale—auto-calibration and the absolute zero associated with it—allows for a simple presentation of dynamic and structural symmetry (reflecting the state of balance) and asymmetry phenomena (reflecting the state of dysfunction). It is also worth to notice the advantages lying in the practical aspect, namely, great ease to compare or even convert data from measurement systems based on the interval scales. For this purpose, the system will be equipped with an interface, allowing for a systemic application of data from devices which show results in the interval scales. The adoption of a universal, multiparametric quotient scale to illustrate the results of instrumental research and to associate it with known, clinical criteria of a medical examination may be an introduction to the formulation of more strict definitions of the patient's condition at the stage of diagnosis, the subsequent stages of treatment, as well as the determination of the recovery criteria (in relation to the parameters of the group standard).

Another issue is the need to develop the technical standard of the device and the measuring points, lying on the body's subsequent levels in a standing position—for easier visualization, they were called transverse measurement planes (TMP), with a strict location for each type of measurement (like the location of the electrodes in the ECG). The output standard should contain a set of horizontal measurement planes and a scheme of their vertical correlations. The next part of the implementation is a standard procedure for testing medical devices, which involves, among others, developing a standard of population norms, in the first phase for healthy individuals (including sexual dimorphism and the process of involution in several standard age groups). In the second phase, after selecting diseases of the locomotor system with particularly specific criteria for the clinical diagnosis, they should be expanded by a formal description of symmetry in standard measurement planes of geometric, kinematic, rheological, thermal, electrical, and magnetic features, whose classification (and then unification) will give the chance to formulate a concise statement setting out the state of parametric symmetry (or asymmetry) of patient's organism, as compared to the group norm, pathology pattern of a disease unit, or comparatively, between the initial state and the next stage of the therapy. Following the above in relation to disease units with specific diagnostic criteria, one can think of attributing sets of relevant coefficients of asymmetry to them, thus enriching the diagnostic criteria with a quantitative dimension. A desired, but still unsurpassed, standard to be followed by locomotor organ diagnosticians should be the topographic-parametric formula of registering in ECG, which FUNCTIONS OVER THE SYSTEM AND LANGUAGE BARRIERS WORLDWIDE.

In experimental studies conducted among large groups of students and healthcare professionals, a repeatable and close relationship between structural symmetry coefficients and symmetry of marker dynamics was observed, and above all, with the coefficients of changes in tissue perfusion, their temperature, resistance, and magnetic induction. The high repetition of the symmetry trend observed in subsequent age groups and the diverse asymmetry in the population of patients raises the hope of creating reliable norm standards for involution phenomena and selected types of pathology.

1.1 Searching for one scale for the symmetry phenomena

Presentation of a global, multiparameter body description on a single graph, which allows simultaneous display of coexistence and correlations between the dominant parameters describing patient's condition, on the one hand requires focusing on the clinical phenomena, which are the most important in this case, and, on the other hand, converting all results registered in a variety of disease units—to one, universal scale.

1.2 Measurement and measuring scales

The measurement is a procedure of associating the features of the examined objects with numbers or other symbols according to specific rules for which (1) the relationship between the numbers and features subject to measurement must be exclusive, (2) the rules for assigning number features should be standardized and applied uniformly, and (3) these rules must not change in relation to the objects and time. The next step is scaling, i.e., creating a continuum on which numerically measured object features are placed. This enables measuring, i.e., assigning to each tested object a strictly defined number within the range of the scale used. Among the scaling techniques, there are (1) comparative scales that allow direct comparison of the features of the examined objects in relative categories and are ordinal or rank and (2) noncomparative scales, for which each object is scaled independently of the others in the examined set, while the results are interval or quotient. Despite the benefits of using comparative scales, such as a common reference point and the ease of detecting small differences between the tested objects and a small number of theoretical assumptions, they have significant disadvantages, which include the ordinal nature of the data and the resulting inability to approximate results outside the population of the scaled test objects.

The quotient scale is the most versatile, because it has all the properties of a nominal, ordinal, and interval (fixed unit) scale, allowing for precise calculation of relationships between the scale values, for which all proportional transformations are permitted. The most important of its advantages, not occurring globally in other scales is a non-arbitrary "0" point on the scale, is the so-called absolute zero, giving the possibility of estimating not only the symmetry of parameters measured by a common unit but primarily a comparison (on one frame of reference) of measurement trends, based on units that are significantly different in their specificity, and virtually using all statistical techniques. Quotient scale, as the most versatile, has all the properties of nominal, ordinal, and interval scale, enabling precise calculation of relationships between the scale values, for which all proportional transformations in the form of $y = bx$ (where b is a positive number) are allowed.

Typization is the process of ranking elements of reality into certain types by referring to existing knowledge resources, preferably parametric (Schütz). A specific element of the surrounding space is treated as similar, provided that the construction and parametric criteria typical of the previously created and known set of features are met. This procedure is an expression of the use of economic simplifications in the classification of objects. Getting to know the surrounding space, you first perceive the general features typical of the class of the object and only then focus on the details. A certain analogy, in particular for the creative process, is standardization, i.e., the activity of analyzing and giving new procedures and innovative products' parametric repeatability to ensure compatibility and limit unnecessary diversity. Profiles of standardized products are made public in the form of standards, technical regulations, or recommended parametric standards (e.g., SI).

The idea of typing for a measuring system is reflected in **Figure 1**, where each of the measured parameters is presented as a plane built into the patient's standard silhouette, on which the lateralization coefficient is determined, which is a reflection of the local parametric symmetry determined on the basis of sensor indications applied at standard marker points. The quotient of these values gives a lateralization

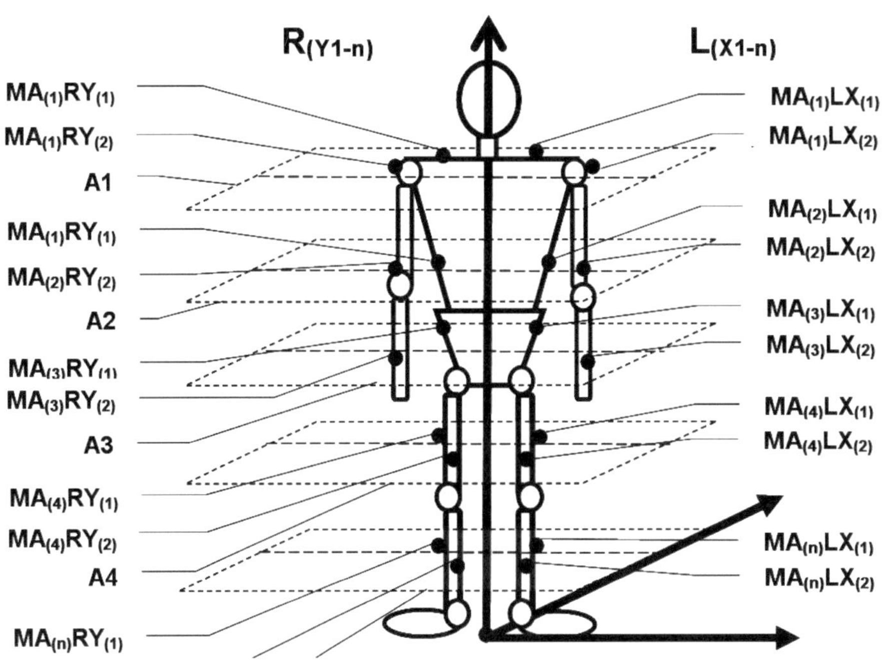

Figure 1.
The layout of marker symmetry on AMSS measurement planes according to the P402850 patent application in the Polish Patent Office, entitled "Method of formulating parametric, diagnostic-therapeutic criteria with an algorithm and device for its implementation".

coefficient, the value of which can depict the symmetry or asymmetry of a given parameter on a common quotient scale. Based on population surveys, mean and reference values as well as the range of standard deviation of results can be determined in subsequent age groups.

1.3 A summary of the project features

While developing the author's concept of anatomical-functional parameterization of symmetry, attention was drawn to the subsequent stages necessary for implementation [5]:

1. Introduction of (employed by us) anatomical standard agreed on transverse measurement planes of the body $A_{(1...n)}$ (**Figure 2**) in order to determine symmetric left-hand $MA_{(1...n)}L_{(x1...n)}$ and right-hand $MA_{(1...n)}P_{(y-...n)}$ markers – their location quite naturally prompts a conversion of any interval parametric scale (exemplary units: V, mA, Ω, um, T) to symmetry and asymmetry indices on a quotient scale, which may reflect symmetry or asymmetry of both, selected features of static images (photography, radiography, and microdensitometry), anatomical structure of twin body organs (joints, bone parts, and points of silhouette contours), dynamic images (quadroscopy and X-ray seriography), and recordable spontaneous (thermography, magnetography, and electrography) and forced emission phenomena (scintigraphy, gammagraphy, and SPECT). It also applies to contact recording of dynamic (accelerometry, gyrometry, and tensometry), electrical (EMG, ENG, EEG, EOG, and PCP), and magnetic phenomena, as well as rheological parameters in symmetrical points of the vascular system (plethysmography, Doppler-USG, angio-CT, and angio-MR).

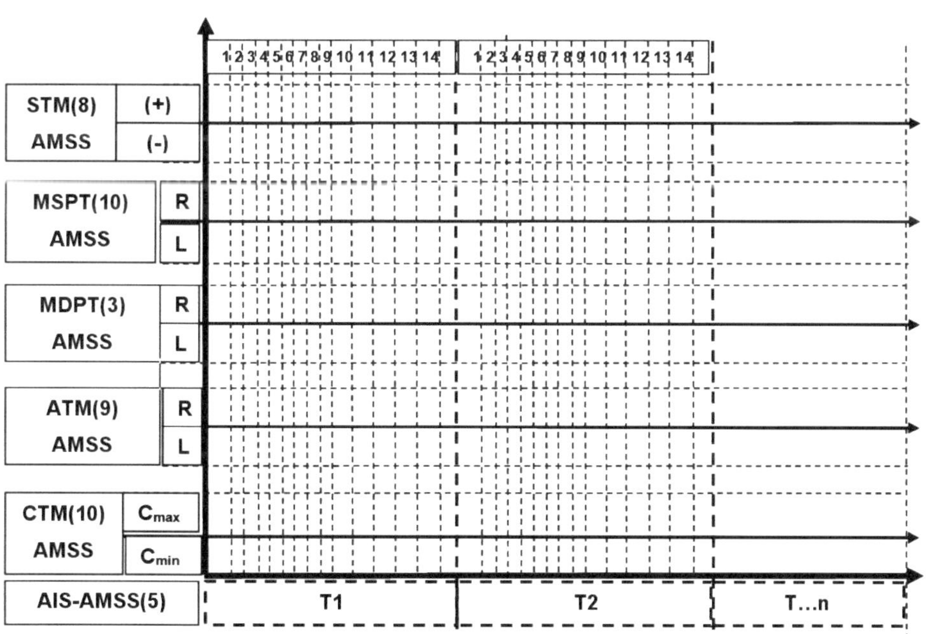

Figure 2.
Exemplary construction of a measurement graph with indication of lateralisation direction (R/L), type of the test (STM—survey tests, MSPT—static planimetric tests, MDPT—dynamic planimetric tests, CTM—biometric tests, ATM—biochemical tests) and sequence of time stamps.

2. In such a constellation, a parametric motion profile (as a single or one of many described functions) of a healthy individual is in fact a sum of (developed over time) parallel sequences of quotients of parameter symmetries ($x \approx 1 \pm 0.1$), calculated individually (as simple or alternating indices) on the basis of data from the left-hand $MA_{(1...n)}L_{(x1...n)}$ and right-hand $MA_{(1...n)}P_{(y-...n)}$ markers located in twin, anatomically characteristic (agreed on) points of the transverse measurement planes of the body $A_{(1...n)}$ (**Figure 2**), while the motion dysfunction in this approach is a sum of parallel, asymmetric sequences ($x < 0.9$ lub $x > 1.1$), for the parameters calculated on the basis of the above standard of markers (**Figure 2**).

3. The results, presented as the (developed over time) quotient scale coefficients, are subject to the laws of statistics, in particular to the correlation calculus, which searches for parametric constellations particularly relevant for dysfunctional syndromes accompanying specific disease syndromes.

2. The associated measurement symmetry system (AMSS)

Detailed assumptions and relevant design parameters of the AMSS original idea have been included in the documentation of patent application P402850, entitled "Method of associating results of several research methods as well as algorithm and device for its implementation" [5].

2.1 Introduction

It is very interesting for any practitioner involved in the diagnosis and treatment of the musculoskeletal system to be able to make a quick and direct measurement of selected characteristics of the geometry of motion vectors and stabilometric posture features against the causative motion parameters (psychometry) and the consequences of movement, among others, contourography, tensometry, accelerometry, electrosensometry, multi-point thermo- and plethysmography, electromyography, nerve conduction, diagnostic imaging parameters, and even psychometric diagnosis of the patient. Implementing this direction of research at LABIOT, at the turn of 1999 and 2000, many clinical tests successfully used a 12-channel system of accelerometer sensors for the analysis of the California gait markers, sensors for the assessment of the respiratory tract, the Lasegue and Yeoman's test, recording the oscillations of selected marker bodies in specific clinical and kinematic states. The advantages of such a configured system have been confirmed in large-scale clinical tests [6, 7]. To meet the need of standard repeatable, optimized measurements of static and dynamic parameters of the human body, and on the basis of them, for an introduction of criterialization of an upright posture and selected dynamic features, it was attempted to group diagnostic tools on a computer platform [8–15], which enables obtaining multidimensional, psychological, planimetric (static), posturometric, and dynamic determinants and thereafter their registration in one database (with a unique ID for each patient). The resulting data is presented on a quotient scale, in the case of biochemical and psychological tests as indicators of the upper or lower limit of the norm, while in the case of planimetric and bioengineering tests as indicators of lateralization. These phenomena have been studied in laboratory and clinically in the succeeding years and were described in the following publications [1, 16–18].

2.2 The essence of the AMSS system

The essence of the AMSS is the method of formulating parametric diagnostic and therapeutic criteria as well as the algorithm and device for its implementation. The purpose of AMSS is to parametrically reflect health and disease (with specified criteria) using the results of simultaneously used research methods that may differ significantly in the dimensions of the physical units used. The method and algorithm give the opportunity to present all test results on a common time axis, preferably on a quotient scale in the form of a graph consisting of dimensionless coefficients related to symmetry, or the minimum or maximum reference value adopted for a specific measuring method (in particular laboratory). This way of recording, processing, and simultaneous presentation of the results of known and commonly used measurement methods opens the way to obtaining a new type of numerical criterion graphs illustrating the state of health and unambiguously pathological criteria. The criterion graph can also be presented in a reduced form using inter-parameter correlation indicators.

The AMSS system, using electronic, algorithmic, and software criteria, allows for the following:

1. Simultaneous, multidimensional recording of biomedical data and their conversion into quotient coefficients, recorded with a time stamp

2. Processing of static, geometric features of body images made with any imaging technique to form symmetry quotient coefficients, recorded with a time stamp

3. Processing of dynamic, geometric features of body images made with any technique to form quotient coefficients associated with symmetry as well as with time marker and path

4. Processing of numerical values for substance concentrations determined "in vivo" and "in vitro" into quotient coefficients that are correlated with the maximum and minimum reference concentration and with a time stamp

5. Processing of numerical results of psychological tests to form quotient coefficients that are correlated with the maximum and minimum value of the reference test and a time stamp

Patient's condition can be registered on the large-format medium (e.g., phone memory card—20 GB) and coupled with a sequence of globally or zone synchronized time stamps (T) as a set of simultaneous, multi-track test results, reflecting the characteristics of the structure and function of the body: (1) quotient indicators (of lateralization), for measurements performed within the markers of symmetrical body parts, and (2) quotient indicators in relation to the maximum or minimum value (for results of questionnaire and laboratory tests).

The purpose of using quotient factors is to create the possibility of the following:

1. Converting the measurement results initially presented using various interval scales to a common denominator, the ratio of the quotient scale

2. Presentation of many, even several dozen parameters of the quotient scale on a common chart using a time stamp to clearly define time relations for interparametric correlation coefficients

3. The application of the numerical values of the above coefficients to form a measurement graph, i.e., a sequence of graphical and parametric symbols that reflect the state of the organism (S), for (n) parameters, recorded with a time stamp (T) (**Figure 2**)

4. Application of the idea of a measuring graph to determine a standard profile graph, for neonatal [GPS (n)], children (2–8 years) [GPS (d)], adolescent (9–14 years) [GPS (m)], juvenile (15–20) [GPS (md)], adult (21–30) [GPS (d1)], adult (31–40) [GPS (d2)], adult (41–50) [GPS (d3)], adult (51–60) [GPS (d4)], involutionary (61–70) [GPS (i1)], involutionary (71–80) [GPS (i2)], senile (81–90) [GPS (s1)], and old age (91–100) [GPS (s2)]

5. Application of the measurement graph idea to determine the sport profile graph [GPS (s1–n)] for athletes in disciplines from 1 to n

6. Application of the measuring graph idea to determine the professional profile graph, in particular for special occupations, e.g., commando [GPZ (ks)], diver [GPZ (n)], pilot [GPZ (p)], and astronaut [GPZ (ka)]

7. Application of the measurement graph idea to reflect the criteria of pathologies of well-verified and characteristic disease entities in order to obtain a dysfunction profile graph [GPD (p1–n)]

8. Determining the progression trend for the dysfunction profile graph (worsening of disease symptoms on the time axis) or regression trend (disappearance of disease symptoms until reaching the standard or sport profile graph in a given age range)

2.3 The structure of the AMSS system

The central element of the system is the CME-AMSS electronic central module, which has at least four groups of I/O ports, enabling data download between peripheral devices and program modules (1–5). The system is controlled by the PP-AMSS measuring software platform implemented in the "DELPHI 7.0 professional" environment, having (1) the BW-SSSP benchmark database; (2) the BWW-AMSS results database, available by patient ID; the SSI artificial intelligence system base-AMSS; as well as a database of tool modules, containing:

1. **TA-AMSS survey test module** is a collection of computer questionnaire tests, in particular psychological and quality of life scales, selected as useful for medical diagnostics, according to the algorithm for creating baseline assessment, assessment during and after treatment. In practice, intelligence tests (Stanford-Binet test, Raven matrices), personality and interests test (Mittenecker-Thoman) and language predispositions, depression tests (BDI, MADRS, HADS, Hamilton), and tree projection test were used.

2. **Planimetric static test module PTS-AMSS** is a computerized package of software tools for automatic planimetric measurements on static files of radiological, scintigraphic, ultrasound, and thermographic images (main location of 2D, 3D points, distance, angles). The use of these easy-to-use measuring tools, and through them the use of a lot of planimetric operations, gives the possibility of creating numerical equivalents of specific elements of the matrix of any image of biological structures. To improve the functionality of the system,

measurements should be made on a matrix of symmetry markers, enabling the acquisition of specific lateralization indicators.

3. **The PTD-AMSS dynamic planimetric testing module** is part of the PP-AMSS software platform intended for the acquisition of measurement data from devices connected to the central CME-SSSP electronic device for extracorporeal measurement of body movements and its anomalies, in particular enabling parametric evaluation related to movement marker drift, recorded as the sum of parallel asymmetric strings marked with a time marker.

4. **The MNA-AMSS analytical tool module** is part of the PP-SSSP software platform intended for the acquisition of measurement data from the CME-SSSP central electronic module connected via an analytical interface. It connects with devices for extracorporeal measurement of sample parameters, mainly body fluids and cells, as well as for measuring physicochemical parameters measured by applicators having direct contact with the examined tissues of the treated patient.

5. **The MTB-AMSS biometric testing module** is part of the PP-AMSS software platform intended for the acquisition of measurement data from dynamic measurement devices connected to the central CME-AMSS electronic module that has a direct contact with the patient's tissues (e.g., plethysmography, pulse oximetry, EEG, ECG, EMG, magnetometry, resistometry and resistive tomography, accelerometry, gyroscopy, impulse perception threshold assessment).

6. **The module of artificial intelligence systems SSI-AMSS** allows, depending on needs, the use of advanced inference systems to perform (a) erosion of the information stream without loss of relevant content, (b) thematic organization of information, (c) diagnostic inference based on decision support systems (hypothesis generators), and (d) inferences based on expert systems and (e) determining the numerical dimension of disease progression and remission trends (under the influence of treatment). A large amount of collected data is connected with the problem of their effective processing, aimed at obtaining cross-sectional knowledge about the studied phenomenon. In medical diagnostics aided by computer methods of data analysis, making the right decision is extremely important because it is about time, life, and costs of patient treatment.

Broadly defined, clinical metrology parameterizes the natural (and test), structural, physicochemical, biomechanical, and behavioral characteristics of the patient in the state of health and disease, giving the possibility of assigning the above features to numerical indicators that facilitate the criteria for health status, specific disease entities, as well as stages of treatment (progression/regression) disease. In our opinion, one of the most important, yet useful, features that facilitate assessment of the clinical state is determining (on the quotient scale) multiparameter lateralization indices, i.e., symmetry disorders in marker points of the twin body structures and presenting the degree of their asymmetry on one, multiaxial graph, where OX axes are time stamp calibrated. According to the authors, this is the first real and moderately expensive method, enabling simultaneous, global multipath view of the patient with the clinical symptomatology of a specific disease syndrome, comparing the subjectively perceived image from the diagnostician with the parameterizing broadband graph of disease asymmetries. It is worth adding that the subjectivity of the patient with whom the interview is collected is partially verified by the TA-AMSS test suite (**Figures 3–7**).

(a) (b)

Figure 3.
Award for the idea and construction scheme of the AMSS at (a) the 7th International Congress of Polish Rehabilitation Society (Łódź 2010) and (b) the gold medal of the 26th Physical Medicine and Rehabilitation Fair in Łódź 2019, for the device - PSB1 posturometer to implement the assumptions from 2010.

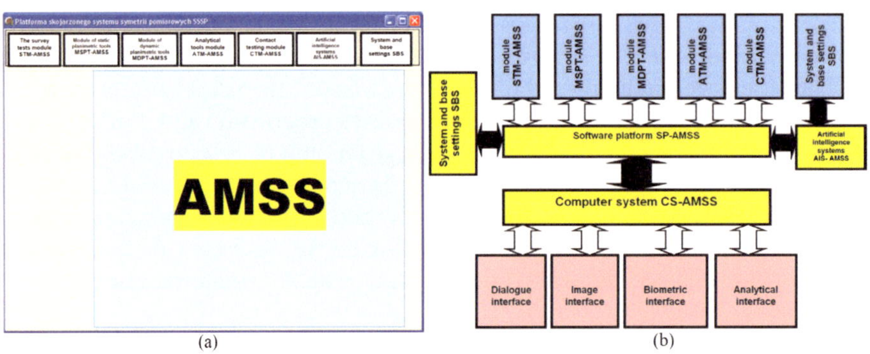

Figure 4.
The main dialogue box of the AMSS system (a) SP-AMSS, connected to a computer system module (b) CS-AMSS, having the dialogue interface, image interface, biometric interface and analytical interface.

2.4 The stream of measurement data

The possibility of using increasingly sophisticated methods for recording parameters of subtle vital functions reveals an increasingly important issue of time interdependence of monitored parameters in clinical metrology. It is not of course of great importance when a single measurement of individual RR, HR, and temperature parameters is made. The possibility of multi-track data recording offered by rapidly developing metrological techniques meant that, together with the sensitivity and scope of monitoring and recording of parallel life functions, the relationship of their instantaneous values with time synchronization on the scale of the whole organism increases. The study of the temporal variability of

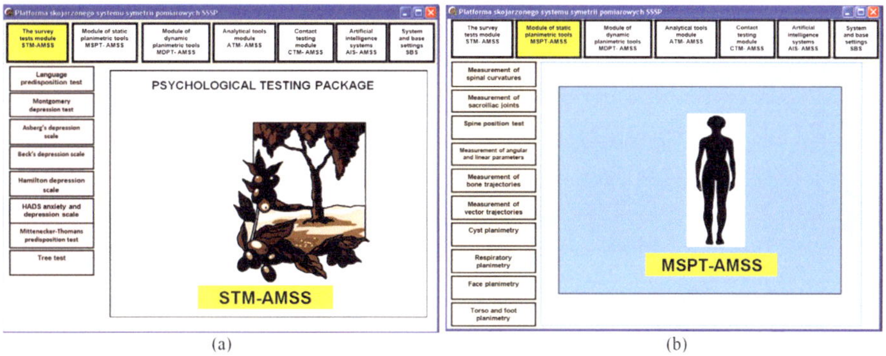

Figure 5.
The software platform SP-AMSS consists of (a) survey tests STM-AMSS (particular psychological and quality of life scales) cooperating with the dialogue interface and the image interface, (b) static planimetric tests MSPT-AMSS (package of utility programs for automatic planimetric measurements on files with static radiological, scintigraphic, ultrasound, and thermographic images) cooperating with the image interface.

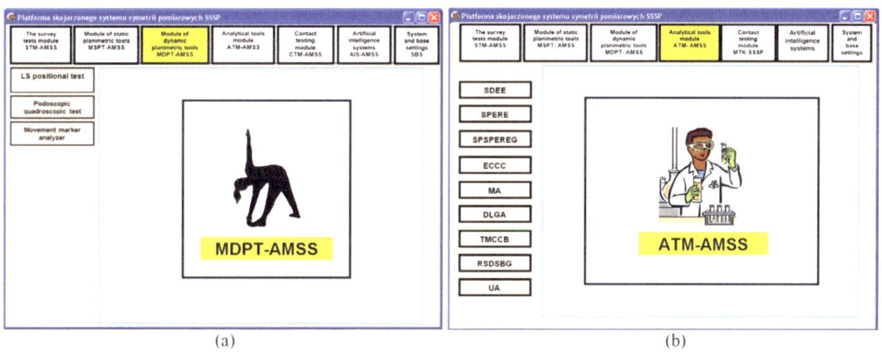

Figure 6.
The software platform SP-AMSS consists of (a) dynamic planimetric tests MDPT-AMSS for the parametric evaluation of movement-associated marker drift, cooperating with the image interface and (b) analytical tools ATM-AMSS (with analytical interface to communicate with devices for extracorporeal measurement of sample parameters, particularly body fluids and cells). Devices in (b): SDEE—separator for dynamic centrifugal and electromagnetic electrochromatography (PL 171643); SPERE—separator for separation of particles by electrophoresis in rotational electromagnetic gradient (PL187203); SPSPEREG—separator for particle separation by phonoinductive electrophoresis in rotational electromagnetic gradient (PL 186372); ECCC—electromagnetic controlled column chromatography (PL 175577); MA—microwave analyzer (PL 194256); DLGA—dynamic liquid and gas analyzer (P416515); TMCCB—transdermal meter for the concentration of chemicals in the blood (W110552); RSDSBG—resonant stimulating and diagnostic system of bone growth (PL 203731); UA—urine analyzer [19, 20].

skin thermoemission, without synchronization with the assessment of blood flow in microcirculation and electrical heart action, may be helpful when the assessment is made by an experienced physician who subconsciously takes into account synchronization patterns. However, in the situation of more and more frequent use of intelligent techniques (e.g., neural networks) and mathematical inference rules in order to generate diagnostic hypotheses by diagnostic support systems, the level of correlation of input determinants depends on the level of correlation of partial determinants and, as a consequence, the reliability of diagnosis. In the introduction, the thesis of the synchronicity of the processes of regulation of homeostasis of the body was put forward. The rapid increase in patient data forces the need for intelligent techniques of data segregation and processing (e.g., rough sets), as well as systems supporting diagnostic decision (expert systems). It should be added, however, that one of the key criteria affecting the quality of

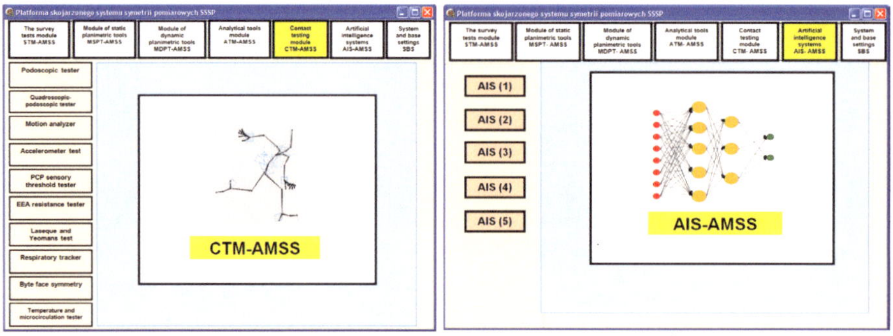

(a) (b)

Figure 7.
The software platform SP-AMSS consists of (a) contact testing module (CTM) for the parametric evaluation of symmetry of parameters of temperature, photoplethysmography, parameters of resistance, impedance, reactance, and microcirculation blood flow, cooperating with the biometric interface and (b) artificial intelligence system (AIS).

the computer system is proper preparation of the input data. Therefore, there is a problem of creating a synchronous measurement and inference system, based on the experience of using the distributed measurement system, which allows the conversion of parameters and their presentation on a common time axis in order to outline interparametric correlations and assess their significance by applying system inference rules. One of the possible solutions to this problem is multipath measurement data acquisition, synchronized by a common time base. The network structure of the system can be easily adapted to the conditions in which patients are diagnosed in several offices, or a series of tests are carried out in different places. Each measurement module is able to attach a time stamp to the data record that is being sent, which makes it possible to determine many factors that depend on it, such as the need for repeated examination or the impact of rehabilitation on the patient's health. The time stamp is an important parameter that significantly affects not only the strength of connections between the test results carried out by individual modules but also the determination of the usefulness of the tests and their validity period. This in turn has a significant impact on economic planning and modification of procedures towards more effective solutions. One of the important elements affecting the results of the system is the proper preparation of the input data. For this purpose, elements of rough set theory will be used. Knowledge is understood here as the ability to classify objects of the studied reality, and its representation uses information systems of which special type is decision boards. One of the purposes of using rough sets is to generate a good, in a sense, universal set of decision rules—sentences of the form: "if trait and (1) = value (1) and (n) = value (n) is decision = value (d)." The creation of decision algorithms involves, among others, problems of knowledge reduction (searching for minimal sets of independent attributes, giving full classification of objects) and searching for dependencies between data. Detection of unnecessary attributes in the decision table will allow obtaining information on the basis of which, for example, you can change the attribute—increasing the time interval or rejecting the attribute. The use of appropriate classification algorithms will allow to reduce the number of objects or change the area of research, e.g., the location of the thermometer. The use of association discovery algorithms will allow, for example, to detect relationships between attributes and thus determine the degree of correlation between them. In the further part of the research program, this will also reveal the relationship between the effectiveness of treatment and the applied medical therapy [21].

2.5 Information structure in the system

The sectoral computerization of a single workplace or even a single healthcare facility becomes a controversial improvement due to the lack of compatibility, and, as a consequence, isolation in the area of information media flows to the traditional part of the system. This creates the need for duplication of media and procedures. A full-size hospital network integrating all workstations in one operating system, which is also used in other facilities, is the only way to effectively use the working time of sector operators and the fast, internal, and external flow of user and feedback information. An important factor determining the effectiveness and efficiency of the system is the central storage of information about operational entities (patients) and the correct solution of the issue of hierarchy of access to read and write data on the part of employees. Cross-department and cross-center patient information flow during consultations, laboratory tests, admission, and discharge should be based on a protocol of operational procedures unified at least at the national level. The first step in this direction, and also a minimal option, is to create an optimal parametric patient graph and an administrative reporting base, whose data, depending on the hierarchy of access (authorization code), can be obtained in the form of a file at any point of the network. The optimal variant assumes the introduction of an interactive archiving procedure, in which a multi-domain measurement file (parametric graph) is created under the patient's ID, which is gradually supplemented during the stay in subsequent diagnostic laboratories, consultation rooms, and hospital wards. The system's layout of file edit fields means that each diagnostic information is entered sporadically from the alphanumeric keyboard and most often applied in digital form, extending the electronically available level of knowledge about the patient's condition. Therefore, information about the patient takes the form of a sequence of parametric graphs on the time axis, which is particularly important when preparing discharge documents, which until now have taken the most time in the practice of a hospital doctor, since their creation required duplication of time using the keyboard in the procedure of manual rewriting of test results from paper medium. When the hospital network has access to the system layout of edit fields and parametric tables of the patient's information file, it automatically enters their content into the discharge document, which reduces the time spent by the doctor on administrative activities by at least 80% [22].

2.6 The role of unified criteria for metrological assessment of the musculoskeletal system

The system association of multidisciplinary packages of measurement files with patient's ID was important but only a first step in the creation of a modern, unified storage system. It should be noted that, this way, objective part of distributed information that so far has been stored on a paper medium was centered and accumulated on electronical medium in a form of numerical laboratory and electrophysiological parameters, which were referred to objective calibration patterns, source files of imaging tests, as well as numerous descriptive information (results of consultations with specialists, descriptions of imaging tests) that do not have its objective calibration pattern. The use of fairly diverse diagnostics in terms of, e.g., locomotor tests and tests of neurophysiological functions used in several wards, creates some communication problems in the field of communicating the assessment of patient's momentary state and even more of clinical results achieved with the use of different types of therapy. This is particularly important when moving a patient from ward to ward or leaving home with the intention of being re-admitted to further therapy at a later date. In the above situations, it is sometimes difficult to objectively determine the initial state and compare it after a period of stay at home or in another ward, and besides, the doctors of subsequent wards are forced to duplicate many initial descriptive activities. In this situation, instrumental and thus quantitative determination of the topography of receptor disorders of the skin, muscle strength, thoracic respiratory path, vector indicators in the Romberg test, California gait indicators of the patient's gait, muscle electromyographic parameters, or nerve conduction, according to the hospital standard, gives the objective component of the state temporary patient, to whom the clinical comment of the doctor is attached. The file is available at any point in the computer network [23].

2.7 Hospital information system

The main goal of the hospital information system (HIS) [22] should be to improve the quality and streamline the patient's treatment process and reduce the operating costs of a hospital. The first SSIs implemented were based on large workstations (mainframes) with uniform software. Currently, the concept of architecture based on network structure and modular software, not necessarily from one supplier, prevails. The network structure, emerging databases, and connection with other hospitals and scientific institutions result in an increase in the quality of medical services provided. The use of computers with a properly designed information system contributes to the elimination of a significant part of errors and mistakes, completes and unifies the data, facilitates faster access to data, allows more efficient use of equipment, and shortens the patient's waiting time for admission to the appropriate department and effective treatment. The exchange of experiences (in particular those described parametrically) between medical personnel of various departments and environments allows for the development of standards in routine clinical management and in unusual situations. The reference structure of the hospital information system should include:

A system of **data presentation** enabling the standardization of patient data and, consequently, sorting, filtering, and classification (e.g., for the purposes of statistics and epidemiology). An important feature is data availability 24 h a day, 7 days a week from any terminal of the network. This postulated computer patient card should contain (in addition to a record of personal information) information about immunizations, allergies to medications, laboratory and imaging tests, opinions of specialists, a list of possible problems and complications, a plan of visits and consultations, issued medical leaves, information on surgeries, prescription medicines, and others. This list can of course be modified depending on the unit. The main elements are as follows: (1) **Knowledge**—a connection through the Internet with other centers. The introduction of the telemedicine techniques—video consultations, transmissions showing the course of innovative surgical procedures and access to the resource databases (MEDLINE), disease codes and procedures, and literature—facilitates the process of diagnosis and treatment. It is necessary here to implement appropriate mechanisms and tools for defining queries and search the available resources. (2) **Communication**—cheap, quick, and easy medium for the physicians to communicate with each other or with patients. Great opportunities are given by the e-mail: e-mail, group mailing lists, SMS technology in mobile phones, videoconferencing, and video consultations. (3) **Alarms, reminders, and suggestions**—automatically generated messages when selected parameters of patient care do not meet the standards. It is a kind of a quality control system, difficult to achieve with other methods. The final decision to take a given action rests entirely with a licensed physician.

2.8 The network structure

At the moment, only systems using Web technologies are being considered [22] as the best solution from an economic and ergonomic point of view, also taking into account the rapid development of the Internet and of related, so-called e-services. Already well mastered LAN and WAN technologies may be used for this purpose. Implemented software must be running in terminal emulation mode or in application mode client–server. A correct and strict control of access to the information is indispensable—mechanisms for logging into the system, password protection, supervised connections (scripting), as well as coding and encryption procedure of data meant to be transmitted. It is important due to the Personal Data Protection Act—L.G. No. 133, item. 833 of 1997. It is also important to ensure compatibility with various hardware platforms and operating systems (Windows 9x, Windows NT, Macintosh, UNIX). Thus, we need to provide multiple ways of presentation that provide for the most popular types of terminals: IBM3270, VT100, MS Windows, X-Windows. Another essential element in the architecture of the proposed information system must be taking into account various network protocols for the transmission and exchange of data, TCP/IP, IPX/SPX, and HL-7, as well as implementing special dictionary translator interface, responsible for the correct interpretation of information passing between the applications of the system. A conversion of data and codes of different applications is usually needed to create a common representation. Usually, such conversion is carried out "on the fly," using appropriate interfaces. To conclude, we also need to take care of a friendly and easy-to-use user interface.

2.9 The elements of artificial intelligence

Information theory (by Shannon) [22] refers to a sequence of events, to the occurrence of which a certain probability can be attributed. Such an approach means that both the amount of information that implies an event and the total amount of information in a particular series of events can be specified. Information is, therefore, the opposite of probability. Repeating information in the transmission (redundancy) reduces the transmission efficiency of the communication system (doubling the sequence) but increases immunity for disruptions in the information canal (loss compensation).

The decision support system is an organized set of experts, procedures, and technical equipment used to support the effectiveness of the decision-making process. Its task is to improve the quality and efficiency of the decision-making process by simultaneous access to knowledge with a structured access path, as well as to the data with no such structure. Thus, the decision support system is a way to make two kinds of system resources available to the user: the source data bank and the bank of operational data processing methods. The computer decision support system is not able to replace human in any part of his activity; thus, it functions as (1) an instrument supporting the memory and tool-making capabilities of the decision-maker; (2) an incentive (or even a push factor) for precision and consistency in formulating research questions, method of demanding, and searching for the scope and criteria for making optimization; and (3) a facilitator for groups of decision-makers (in particular) operating at lower levels of the management structure, allowing to make informed decisions without the need to have confidential knowledge about the whole system. The development of decision support systems is carried out in the following directions: (1) interdisciplinary modernization of system methods by employing specialists not only in narrow specialties but above all in theory of decision-making, management, operation research, cognitive psychology, fuzzy logic, and mathematical modeling; (2) the broadly defined typing of information technology, the development of which moves toward electronic data collection and conversion of existing archives; and (3) development of technology and methods of communication with the electronic archives. Features of the decision support systems (DSS) include (1) presenting information in the form that is known to a group of users, (2) selectivity of the information provided, (3) supporting and not replacing the decision-making process of the decision-maker, (4) supporting the structuring of the decision-making process, (5) flexibility and quick response to the user needs, (6) supporting various styles of decision-making, (7) acting in different phases of the decision-making process, and (8) cross-checking of various phases of intertwining decision-making processes realized according to different styles.

An expert system is a technically implemented or software-independent algorithm modeled on the activity of the nervous system that can, on the basis of provided or measurement data, make decisions regarding regulation of the control object in a specified range of prediction. It is the use of computer that solves complex problems requiring expert knowledge. To achieve this objective, the expert system simulates the process of human reasoning, using specific scientific knowledge of nature expressed with mathematical inference rules. An expert system (according to Feigenbaum) is "an intelligent computer program that uses inference procedures for solving problems, which are difficult enough to require significant expertise of the specialists. Knowledge, together with the rules of inference, can be regarded as a model of expertise normally achievable only by top professionals." The following types of DSS are distinguished: (1) **Specialist**—bespoke for specific users, regarding their specific problems and ways of solving them. (2) **System generators**—programs allowing for construction of specialized SWD, quickly and easily, based on the system core, often with no need to consult with industry authorities. (3) **Tools**—software supporting other systems, often added as modules or separate programs (graphics system, database type and method of access, software packages, etc.). Within DSS, there are categories that can be distinguished: (1) **Consulting systems**—presenting solutions to the user that can assess their quality. They are usually collected in order of increasing probability. (2) **Systems making decisions without human control**—they are used for the control of dangerous or microautomatic systems (e.g., insulin pumps). (3) **Criticizing systems**—they verify the problem presented to them in the form of a graph, equation, or diagram, showing a list of discrepancies (with known rules) and a list of suggested corrections. The result of an expert system may be the following: (1) **Diagnosis**—assessment of the current state of the object, based on the data processed on the basis of the installed inference rules. (2) **Prognosis**—prediction of future states of the data processed on the basis of the installed inference rules. (3) **Plan**—a sequential-structural description of the state to which the system aims at in a given operating cycle.

Expert systems (ES) are an evolutionarily younger product of the development of knowledge engineering, often based on the existing knowledge support systems. Both systems operate with the use of (1) artificial intelligence, (2) design methods, and (3) functional fusion of systems. The backbone of artificial intelligence can be used to build both DSS and ES. The main directions of the application are (1) model knowledge bases in the range representation codes especially for specific disease entities, access paths, generators of convergent, and alternative and contradictory definitions and (2) construction of sophisticated models of the simplest access and data presentation interfaces, with a tendency to engage in dialog in natural language (according to the principle of "the advanced background technology") [23].

In the large system of spatial regulation of the whole organism, there is a hierarchical-parallel organization of the interdependence of individual centers and circuits. It consists in the fact that the regulatory influence of parent structures on a number of processes at the lowest level is limited in the possibilities of changing local parameters only to the optimal values, which fall within the range of autonomous function of the subsystems. Principles of intersystemic cooperation shaped this way cause, that in moments of special mobilization of the entire body, it is not at the expense one organ or group of cells only, but each of the systems contributes to homeostasis with its "regulatory surplus" that exists around its optimum of functioning. The disease process is based on local or generalized overrun beyond this principle. Creating models simulating the functions of living organisms should not be mere mind gymnastics or a test of skills in practical application of the possibilities of mathematics, electronics, and software, but a practical and functional facility solving specific problems in various fields of science. It is necessary to realize that in modern science, there is a number of excellent diagnostic methods used at the most basic level, without mathematical apparatus that guarantees high specificity and accuracy. A key task for an IT specialist in biology and medicine is the use of digital machines for the processing and description of signals or images of biological objects. Combining the power of the mathematical apparatus with an instant access to data library containing expert knowledge, a significant deepening of the level of problem understanding is obtained, even at the stage of preliminary data analysis. In this new situation, a doctor or biologist will have knowledge, normally reserved for highly specialized centers, and the practical consequences of this fact should not be long to wait for.

However, in order to use digital computational technique in humans or other living organisms, there is a need to "translate" the axial physiological functions of the body, both in the state of health and sickness, into the strict language of structural algorithms that can be identified with the "disease models," compiled then with higher-order language to the level of digital functions. The most valuable feature of the artificial neural networks is their ability to process information in a parallel way, as well as the learning process, that replaces traditional programming. The operation program of the network is included in the topology of connections and their parameters, that is to say weights. Solutions to the tasks the network should solve are presented by means of the output signals. The most important problem during the construction of an artificial neural network is the selection of the network architecture to a particular problem. The choice of a neuron model and the method of teaching the network are also important. The best solution for the classification tasks is one-way network, while, for example, for complex optimization structures, a feedback network. Learning is a process that leads to consolidation of specific behaviors, based on experience. From the technical point of view, the problem boils down to determining the product of the scalar signal vector and the instantaneous weight vector of a discrete network element that can operate in a time or space pattern. In the artificial neural networks, learning is a process where each stage is usually visible as a cause-effect relationship. Teaching the network means forcing specific response to the input signals given. In artificial neural network applications, supervised learning and indirect supervision learning can be distinguished. Teacher-assisted training of the network requires preparing an appropriate set of input and output data, i.e., the training set. During learning, each cycle checks whether the network response is as expected—if not, then the weight change process takes place. Teaching lasts until all responses to the learning set data are correct. They are considered to be such, when the difference between the output of the network and the expected signal is less than the established maximum error.

2.10 Clinical implementations of AMSS system elements

I introduced lateral coefficient methodology in medical diagnostics in 2000, initially to the description of the Lasegue test [24], later to assess goniometric techniques, X-ray planimetry, and podoscope planimetry [25]. At the turn of 2003–2005, the idea of the lateralization indicators developed in the direction of multipath symmetries [15]. Since that time, the idea of multipath symmetries was successfully tested clinically on medium groups of patients [26, 27] and large groups of patients [6]. A very important consequence of this completely new approach to the diagnosis was making an important discovery and acquiring clinical evidence that together with asymmetric scheme of long-term physical adaptation of the organism in the course of pain syndromes, bone remodeling occurs in the twin supporting elements of the bone structure, which can be described using lateralization indicators [28].

A summary of the above idea was presented in 2010, in Łódź, at the 7th International Congress of the Polish Rehabilitation Society, in the form of paper entitled "A system of associated measurement symmetries as an EBM tool in standardizing the diagnosis and surveillance of the musculoskeletal system treatment," and then awarded by the Scientific Committee of the Congress. This paper contains the idea and the general modular design of associated measurement symmetry system (AMSS). This issue is discussed in more detail in two chapters of the post-conference monograph [29–31]. Techniques of describing patient's clinical condition (taken from the SSSP system) were also used in a large clinical study, published as a book chapter in the United States [1].

2.11 The role of the AMSS system in anthropometry

The primary goal of the AMSS system, especially in the context of anthropometry, is to present vector biomechanics parameters as part of a time-oriented parametric graph. It is worth recalling and emphasizing that the parametric graph contains the following:

1. Subject's emotional profile, determined using survey tests (intelligence-abilities-depression) and planimetric silhouette tests (parents, family house, tree) (TA-AMSS), approximating more important features of the emotional state, which may affect the qualitative dimension of intentional movement creation.

2. A biochemical profile of the subject (MNA-AMSS), which could have a modifying influence on the qualitative selection of intentional engram of movement creation (e.g., thyroid or adrenal gland dysfunction), the functioning of the neural network in which the engram will operate, the functioning of the motor neurons and pyramidal tract, the functioning of the neuromuscular feedback (motor plates), and the functioning of the muscles.

3. A profile of the subject's static structural symmetry, determined by the module of planimetric static tests PTS-AMSS, which contains a set of utility programs for automatic planimetric measurements in static radiological, scintigraphic, ultrasound, and thermographic images (mainly symmetry, distances, and angles). The application of these easy-to-use planimetric tools gives the ability to create numerical symmetry equivalents for specific pairs of markers on the images of the subject's body structures. To improve the system's functionality, measurements should be performed on a standard matrix of symmetry markers, enabling the acquisition of specific lateralization indicators and above all free, inter-center exchange of information while creating a knowledge capital about the patient, with no unnecessary duplication of procedures in subsequent centers.

4. A telemetric profile of the dynamic, contour-structural symmetry of the subject, determined by the module of dynamic planimetric tests (PTD-AMSS), containing devices for telemetry measuring the symmetry of body movements and its anomalies, in particular enabling parametric evaluation of drift-related

asymmetry indicators, recorded as the sum of parallel, parametric asymmetric strings, marked with a time marker.

5. A contact profile of the subject's dynamic symmetry, determined using the module of biometric tests (MTB-AMSS), equipped with measuring devices having direct contact with the patient's tissues (e.g., tensometry, plethysmography, pulse oximetry, EEG, ECG, EMG, magnetometry, resistometry and resistive tomography, accelerometry, gyroscopy, assessment of the impulse perception threshold). It collects information about the functional symmetry of motion (muscle) effectors, the functional symmetry of peripheral effector control (nerves), and the functional symmetry of effector supply (blood vessels).

6. A system of time stamps.

The multitude of parameters obtained from the AMSS system will, in the first phase, hinder the orientation, even for an experienced doctor who is not a specialist in anthropomotorics and psychomotorics. This problem can be solved by the module of artificial intelligence system SSI-AMSS which (depending on the needs) allows for the use of advanced intelligent tools to perform (a) the erosion of the information stream without the loss of significant content, (b) thematic organization of information, (c) diagnostic inference based on decision support systems (generators of hypotheses), and (d) inference based on expert systems and (e) determine the numerical size of disease progression and remission trends (resulting from the treatment). A large amount of collected data is related to the problem of its efficient processing, aimed at obtaining cross-sectional knowledge about the patient. In medical diagnosis supported with computer methods of data analysis, making the right decision is crucial as it determines the time of making diagnosis and treatment initiation and, consequently, patient's life and the costs of treatment.

2.12 Further development plans for the AMSS system

Further work (modeling on the ECG cardiological standards) will aim at the medical community's interest in creating a friendly and comprehensive numerical standard of musculoskeletal system symmetry (SMSS), determining symmetry coefficients (or asymmetries) according to topographic patterns of the osteoarticular, muscular, nervous, and circulatory systems [32]. The purpose of implementing the new system is to provide a new, more universal tool that can be used to create a protocol for inter-center exchange of information. The program plan includes the following:

1. Improvement, unification, and typization of the marker matrix (MM), a system for electronic registering of data and software.

2. Development of the diagnostic screening package (DSP), which is a set of optimal topographic and parametric patterns (TPP) characterizing the state of the organism.

3. Establishing a set of physiological patterns (PP) that characterize the normal state of the body and will function in the information database as a frame of reference. This data will be collected after applying the diagnostic screening package (DSP) in studies on subsequent large age groups of healthy individuals, thus reflecting a natural phenomenon of parametric involution.

4. Optimization of the qualitative selection of the topographic and parametric patterns (TPP) in specific disease units, in order to be able to repeatably present them as indicators of symmetry (or asymmetry) on the quotient scale.

5. Beginning to work on the base of pathology patterns (WP), through the use of the diagnostic screening package (DSP) in studies on selected groups of patients for whom the selection process was carried out on the basis of highly specific diagnostic criteria.

6. Multi-center clinical tests, leading to the parametric unification of physiological (PP) and pathology patterns (WP).

7. Reconstruction of electronic devices, software, and clinical criteria—based on feedback.

8. Multi-center unification and typification of the system and procedures.

Author details

Andrzej Jan Dyszkiewicz[1,2*] and Diana Hruby[1,3]

1 LABIOT Laboratory of Biotechnology, Cieszyn, Poland

2 Higher Technical School in Katowice, Katowice, Poland

3 Academy of Physical Education, Katowice, Poland

*Address all correspondence to: andrzej@labio.pl

References

[1] Dyszkiewicz A. Chapter 2: Local use of electrophonophoresis in the treatment of patients with immunologically based neuro-vertebral conflicts. In: Montenegro ML, editor. Pelvic Pain: Causes, Symptoms and Treatments. Nova Science Publ Inc; 2012. pp. 45-76

[2] Dyszkiewicz A, Nowak E, Szczegielniak J, Szurmak T. Planimetric parameters of the tree test in a quantitative evaluation of the susceptibility to examination stress and acoustic stress. In: Advances in Soft Computing, Information Technologies in Biomedicine. Springer; 2012. pp. 245-258

[3] Dyszkiewicz A, Kępiński P, Połeć P, Chachulski D, Szczegielniak J. Komputerowa ocena efektów leczenia bólu wodcinku lędźwiowym kręgosłupa. Pomiary, Automatyka, Kontrola. 2012;**4**(58):339-343

[4] Dyszkiewicz A, Kępiński P, Połeć P, Chachulski D, Nowak E. Pomiarowe efekty zastosownia galwanomagnetostymulacji w leczeniu szyjno-barkowych zespołów bólowych. Pomiary, Automatyka, Kontrola. 2013;**3**:196-199

[5] Dyszkiewicz A. Sposób formułowania parametrycznych kryteriów diagnostyczno-terapeutycznych oraz algorytm iurządzenie do jego realizacji. Zgłoszenie. 2013:P402850

[6] Dyszkiewicz A. Wpływ Elektrofonoforezy Levamisolu na Wyznaczniki Symetrii Ciała i Stężenie Kompleksów Immunologicznych w Wybranych Zespołach Bólowych Okolicy Lędźwiowo-Krzyżowej Kręgosłupa. Monograficzna Pracanaukowa Nr. 2353. Katowice: Wydana Przez Uniwersytet Śląski; 2005

[7] Dyszkiewicz A, Opara J. Znaczenie posturometrii i stabilografii w rehabilitacji i zapobieganiu upadkom u osób poudarze mózgu. In: Zeszyty Metodyczno-Naukowe: Nowoczesna Rehabilitacja po Udarze Mózgu. Katowice: AWF; 2006. pp. 127-140

[8] Dyszkiewicz A et al. The use of a digital goniometer to assess the angle of joint mobility. Acta Biooptica et Informatica Medica. 1999;**5**:81-85

[9] Dyszkiewicz A et al. Accelerometric diagnosis of the balance organ and treatment monitoring. Technical Problems in Medicine. 1999;**1-4**:17-21

[10] Dyszkiewicz A, Dyszkiewicz U, et al. Psychometric analysis of personality and talents in the interactive process of dialogue with the computer. Cognitive Science and Media in Education. 2000;**1-2**:209-220

[11] Dyszkiewicz A et al. Diagnosis of lumbar spine based on static-dynamic, motor radiological patterns. Physiotherapy. 2004;**12**(3):21-28

[12] Dyszkiewicz A, Gaździk T. The use of immunomodulatory drugs in the treatment of lumbar disc-radicular conflicts of two-track symptomatology. Journal of Orthopaedics, Trauma and Rehabilitation. 2004;**6**(2):190-197

[13] Dyszkiewicz A, Kucharz EJ, Rumanowski M. Biomechanical aspects of the axial function of the spine in the human body. Physiotherapy. 2006;**14**(4):79-92

[14] Dyszkiewicz A, Opara J, Gaździk S. Monitoring the treatment of pain syndromes of the lumbar spine with the use of non-steroidal anti-inflammatory drug and gel based on natural-essential-oil component. Journal of Orthopaedics, Trauma and Rehabilitation. 2006;**2**(6):210-218

[15] Zielosko B, Dyszkiewicz A. Intelligent Medical Systems on internet technologies platforms. In: Advances in Soft Computing, Monitoring, Security and Rescue Techniques in Multiagent Systems. Berlin/Heidelberg/New York: Springer; 2005. pp. 589-594. ISBN: 3-540-23245-1

[16] Dyszkiewicz A, Kępiński P, Połeć P, Chachulski D, Szczegielniak J. Computer evaluation of the effects of pain treatment in the lumbar spine. Pomiary, Automatyka, Kontrola. 2012;**4**(58):339-343

[17] Dyszkiewicz A, Kępiński P, Połeć P, Chachulski D, Nowak E. Measuring the effects of using galvanomagnetostimulation in the treatment of cervicobrachial pain syndromes. Pomiary, Automatyka, Kontrola. 2013;**3**:196-199

[18] Dyszkiewicz A, Kępiński P, Połeć P, Chachulski D, Nowak E. Controlling dynamics of a collimator with ferrofluid in the formation of the spectrum of diagnostic beam and reduction of defects in X-ray images. Pomiary, Automatyka, Kontrola. 2014;**4**:207-211

[19] Reh ref moč měch úraz míchy s použ elektron reg průt. Reh fizik lek. Praha. 2005;**1**:25-35

[20] Dyszkiewicz A, Szczegielniak J, Opara J, Chachulski D, Połeć D, Zajdel J. Electronic urine flow reg in post-trauma sphincter-detrusor reflex atrophy of the urine bladder. Rehabilitation Medicine. 2008;**47**:129-130

[21] Dyszkiewicz A, Gaździk T. Monitoring of the sacroiliac joints inflammation and their treatment by using electrophonophoresis with levamisole and butapirasolum. Annals Academy of Medicine Siles, Katowice. 2002;**47**:226-239

[22] Duch W, Korbicz J, Rutkowsko L, Tadeusiewicz R. Sieci neuronowe. In: Nałęcz M, editor. Biocybernetyka I Inżynieria Biomedyczna Tom 6. EXIT; 2000

[23] Hampton AN, Bossaerts P, Doherty JP. Neural correlates of mentalizing-related computations during strategic interactions in humans. Proceedings of the National Academy of Sciences of the United States of America. 2008;**105**:6741-6746

[24] Dyszkiewicz A, Gaździk T. Treatment and monitoring of immune therapy of arthritis of the knee using levamisole and butapirasolum administered with transdermal electrophonophoresis system. Annals Academy of Medicine Sile, Katowice. 2002;**47**:240-251

[25] Grush R. The emulation theory of representation: Motor control imagery and perception. The Behavioral and Brain Sciences. 2003;**27**:377-442

[26] Dyszkiewicz A, Kucharz EJ, Rumanowski M. Application of electrophonophoresis in the treatment of periarticular inflammation of the shoulder joint. Physiotherapy. 2004;**3**:67-72

[27] Dyszkiewicz A, Gaździk T. The use of immunomodulatory drugs in the treatment of lumbar disc-redicular conflict with two-track symptomatology. Journal of Orthopaedics, Trauma and Rehabilitation. 2004;**6**(2):190-197

[28] Dyszkiewicz A. The Impact of Levamisole Electrophonophoresis on the Determinants of the Body Symmetry and Concentration of Immune Complexes in Selected Pain Syndromes in Lumbosacral Spine. Monographic Thesis No. 2353. Katowice: University of Silesia; 2005

[29] Dyszkiewicz A, Połeć P, Zajdel J, Chachulski D. Specific evaluation of

pelvic radiograms and hip BMD in structural scoliosis reflectorica and reactive pain conditions of the backbone. In: Advances in Soft Computing, Information Technologies in Biomedicine. Springer; 2010. pp. 491-450

[30] Dyszkiewicz A. Standardization of the motor organ diagnosis. In: Klukowski Krzysztof Sports Medicine. Med Trib; 2012. pp. 51-55

[31] Dyszkiewicz A. Application prospects of the system of associated measurement symmetries SSSP in rehabilitation. In: Klukowski Krzysztof Sports Medicine. Med Trib; 2012. pp. 56-61

[32] Kącki E, Kulikowski J, Nowakowski A, Waniewski E. Systemy komputerowe i teleinformatyczne w służbie zdrowia. In: Nałęcz M, editor. Biocybernetyka I Inżynieria Biomedyczna. EXIT; 2000

Chapter 4

Biomechanics as an Element of the Motion Clinimetry System

Andrzej Jan Dyszkiewicz and Diana Hruby

Abstract

The study highlights the great progress in medicine, currently capable of a criterial, qualitative diagnosis of an increasing number of primary and secondary diseases in the musculoskeletal system, regardless persistent obstacles to a credible, systemic, and quantitative evaluation of the extent of existing motion dysfunctions, as well as subjective dimension of patient's suffering. It is worth to add that only parametric estimation of a qualitative dysfunction profile makes it possible to reliably monitor treatment efficiency and forecast the level of health damage after its termination. The essence of biomechanics, understood as a science describing internal and external forces' vectors, which determine specific, dynamic motion models (especially for balance and gait), has been presented in the study. Special attention has been given to anthropomotorics and psychomotorics, which give a broader context to motion's driving phenomena and consequences, thus offering a variety of new parameters that have not been considered in close relation to motion so far. While developing symmetry concept, it was pointed out that dysfunction profile comprises of sequences of parametric asymmetries registered in twin body markers.

Keywords: essence of biomechanics, anthropomotorics, psychomotorics, concept symmetry of body parameters, clinimetry, registers static and movement parameters

1. The essence of biomechanics

Biomechanics is an element of clinimetry, which registers movement parameters and analyzes complex inter-parameter phenomena, mainly to an extent to which they reflect the impact of internal and external mechanical forces. There are two main directions in observation, namely, retaining balance and moving body (or its parts) within Earth's gravitational field. In biomechanics, there is a slightly archaic and misleading but still bounding division for statics (applicable exclusively to the rigid bodies) and dynamics, which seems to overlook the fact that maintaining a seemingly static posture results from a conjunction of micromovements that project the center of gravity into a closed curve of critical supporting plane. A distinguishing criterion between a seemingly static and dynamic behavior of a living organism lies exclusively in technical capability to register micromovements as well as macroscopically visible movements [1]. In classic approach, dynamics can be divided into kinematics, describing movement geometry, and kinetics, describing movement's driving forces. In the state of static balance, as well as in any body's movement, the impact of certain forces can be evidenced, relying on well-known physical principles, especially vector geometry [2]. However, is the location of the center of gravity and inertia, balance, and vector calculus of marker points of the human body equal and repeatable as in the case of the description of the twin points of a rigid body, e.g., a stone sculpture or a polyester 3D print of a human body? Are there in practice differences between the physical characteristics of the behavior of marker points of rigid bodies and living organisms composed of multi-compartment, complex rigid systems with a specified number of degrees of freedom (bones and joints) and viscoelastic kinetic-buffering systems, with stiffness varying depending on the concentration and the direction of ion migration and the current pattern of reflex muscle tone stimulated by the extrapyramidal system? Is the repeatability of the vector calculus (for two bodies of identical mass) carried out after the action of force with constant parameters different for the marker point, which is part of the homogeneous structure of the rigid body, compared to the description of the marker of a multi-compartment rigid-viscoelastic system? What is the parametric drift of the viscoelastic system, the rigidity of which is interactively modified by a vector of information flowing from the inside of the body and the environment through receptors to the CNS and then by extrapyramidal pathways of the spinal cord to the muscles?

Biomechanics seeks to answer many of these questions pragmatically, by means of creating parametric sets for functions of motion markers in healthy individuals in various age groups, e.g., in the development period (evolution patterns) and in the old-age period (involution patterns) [3]. Clinical biomechanics covers sets of parametric differences in motion dysfunctions caused by a number of disease processes [4]. When applying mathematical formulas for describing rigid bodies in clinical biomechanics, it should be noted that motion is an effect of not only external and internal forces but also other internal factors, such as information vector, which is not usually directly associated with the causative aspect of motion, or else there is no relevant mathematical formula to describe it yet [5]. An important element of basic research on human motility is modeling, where body movement is treated as an effect of a control process, in which the controlled object is a simplified, mechanical model of a rigid-viscoelastic anatomical structure, and the controlling element is a regulator, which reflects the functions of sensorimotor part of the nervous system. Built in this way, a motion model is subjected to the computer simulation and later to a correlation analysis with real data acquired from sensors installed on volunteer's body, in order to modify the model. Therefore, when planning to improve biomechanical models, it is worth thinking about adding other factors to the vector analysis that may play a role in increasing the reliability and repeatability of the description of the movement of biological organisms [6]. So far, none of the existing research patterns were able to meet the strict consistency criteria in science. Every theory requires to be confirmed by experiment, and attractive, probabilistic models, which were constructed so far, operate in isolation from the actual empirical data and have moderate practical significance [7, 8].

Distinguishing anthropomotorics as a science was an attempt to find a compromise between the inductive and deductive approach in studies on human movement—it seeks to objectively examine the phenomenon of motility in all its aspects in order to provide the specialists in various fields with one uniform pattern, which would correspond with results of the real measurements. One of the most important issues in the field of anthropomotorics is an attempt to cybernetically define the characteristics of the control system (regulator) included in the model above—its functioning defines complex paths of interactions between sensors and motion effectors (muscles) and moving rigid modules (bones), connected by ties (joints) with a certain number of degrees of freedom [9–11]. Research on the complex functions of the motion regulator funded a new discipline—psychomotorics—which defines the role of senses, largely proprioceptive (tensometric strain gauges) and labyrinth (accelerometer-gyroscope) sensors, and the use of information obtained through them to modulate the course of movement in terms of purpose function (verified by the telereceptors) [12–14]. It is worth to remember that psychopathological phenomena, and even transient emotional states that modify one's personality profile, influence the motion control, causing visible amplitude-interval changes, e.g., in Californian motion determinants. This is why it seems very legitimate for various authors to suggest the necessity of testing the stability of this biological regulator also in psychological aspect, by means of parallel use of psychometrical test and movement metrology [15–17].

2. Structural and functional symmetry of the body

Majority of living organisms that have evolved in the gravitational field of the Earth, except for a large part of protozoa, known as non-axial, developed at least one symmetry axis and usually a symmetrical arrangement of many internal organs and body parts. The symmetry of the body is, therefore, a basic and quite specific feature of the structure, defining the distribution of its parts in relation to the adopted axis or plane. Using the symmetry criterion, the organisms can be divided into:

1. Radial symmetry organisms having a large number of symmetry planes, running through the body along a single symmetry axis. According to the number of rays, they can have either a biradial symmetry (for two radii and two planes of symmetry) or a quadriradial symmetry (the four radii and four planes of symmetry). Examples include vascular plants, sponges, polyps, jellyfish, coelenterata, polychaetes, and echinoderms [18].

2. Organisms with bilateral symmetry, defined by the plane running along the main (long) body axis, which divides it into two parts, the right and the left. Bilateral symmetry is a construction plan both for animal (e.g., amphibians, reptiles, birds, mammals) and plant organisms (uni- and multicellular) [19].

Each of the organ pairs existing in the living organism (with bilateral symmetry), despite the high level of structural distinctiveness, functions in a substrate-coupled manner with odd central organs (heart, liver) by means of mirror elements of the vascular system and information coupling with the brain via the peripheral nervous system, seeking to maintain their structural and functional symmetry. It has been proved that information and ion-molecular regulatory phenomena that penetrate from the microscopic to macroscopic scale, by maintaining functional symmetry, also tend to maintain structural symmetry that can be found in shape geometry at all levels of twin organs (e.g., limbs). These features can be described by the laws of formal logic and reflected in a model as structural-functional symmetries of parameters, not only amplitude-angle but also thermo-emissive, rheological, magnetometric, resistive, and electrometric. As mentioned above, one example can be balance in a seemingly static standing posture, which is a characteristic feature of human motor activity, defined as the body's ability to maintain body position without assistance, allowing to maintain this state during and after specific activities [20]. Balance in its essence is therefore a dynamic act of mobility, whose stability is conditioned by the efficiency of spatial neuromuscular coordination in the field of statokinetic microreflexes. Its expected result is the projection of the body's gravity center into a closed critical curve of the supporting plane, thus maintaining the standing posture. This feature is one of the external manifestations of the homeostasis of the body, which tends to keep symmetry of function in the twin body organs. Thus, both the statokinetic micromovements that condition maintaining a relaxed standing posture and more easily discernible features of the symmetry of gait phases indicate the efficiency of the osteoarticular, muscular, and nervous systems. Another variable characteristic of balance is stability, that is, the ability to restore the position of the body in space, that was lost as a result of external forces or one's own physical activity.

3. Anatomical and functional outline of balance phenomena

There are three main groups of receptors adapted to the perception of the data necessary for the smooth functioning of the regulator in control of body balance (seen as an irregular body), namely, the vestibular system, the proprioceptor system, and eyesight and hearing under certain conditions. *Signals generated by these structures are a source of information about the position of the body and its orientation relative to external and internal reference systems* [21].

In the process of postural balance control, human brain creates an internal reference system in the form of a symmetric, proprioceptive simulation, which is a dynamic, spatially symmetric strain gauge load distribution model in the body's schemes of postural and motor habits, calibrated by an external reference system and engrams of acquired variants of gravity deviations [22]. This model is referred to as an external visual reference system, which is a dynamic, optical model of the external space, which surrounds the subject [23–25]. Under specific conditions, acoustic model can also be considered [26]. One of the elements of the subject's interaction with an external reference system is the need to determine the relationship of its own dimensions in relation to the marked, repetitive components of the environment. Archaic measurement systems (where, e.g., foot and cubit serve as a unit) were created in this mechanism. Internal and external markers allow to monitor the deviations of the center of gravity and inertia of the body and even its individual parts from the state of balance. Anatomy-wise, eyesight and the vestibular system are organs that monitor the position of the head in space, whereas proprioreceptors and exteroceptors form a network of sensors covering the whole body. Each of these recorders collects a different type of data that individually affects the operation of the regulator and systemic balance control in a specific way [27].

In this context, it is worth to mention the structure of the vestibular system, which is composed of three semicircular canals located in mutually perpendicular planes. On the inner walls of the channels, there are receptors (hair cells) transmitting information on the direction and speed of movement of the head. Other parts of the vestibular system, saccule and utricle, are located at the base of the semicircular canals. From a structural point of view, they are two chambers filled with endolymph, in which the bottom walls have concentrations of hair cells—their cilia are located between the otoliths submerged in endolymph, which are built of crystals of calcium salt crystals. Their function is similar to the seismic mass in accelerometers. Signals from these organs transmit information about the static position of the head in space, first to the spinal cord via the vestibular nuclei. This is followed by the integration of signals from the otoliths, cerebellum, and spinal cord in the lateral vestibular nuclei (Deiter's nuclei), where efferent signal sequences that stimulate α- and β-motor neurons of the spinal cord (in control of muscles) are being generated [28].

As already mentioned, the regulation of spatial orientation, of balance in particular, is also based on the analysis of the image registered by the eye. In this case, the role of the external reference system is performed by the geometrical relations of the spatial location of other objects on the proximal and distal set. In the neural network of the retina, there are specialized groups of neurons that perform loss conversion of the image fragments in order to obtain the vertical and horizontal gradients. These data have a similar role to the signals from the vestibular system [29–34].

The network of proprioception receptors (proprioceptors) is located mainly in collagen bonds connecting rigid modules (bones), i.e., in joint capsules, ligaments, and tendons, as well as in viscoelastic modules (muscles). A network of sensors sensitive to mechanical deformation generates data to create a projection-proprioceptive simulation, and after the confrontation with the model's pattern database—a proprioceptive-cognitive simulation of spatial distribution of the load in the individual body modules, the rigidity of viscoelastic elements and intermodular stress. The conversion and transmission of this data to the dimension of conscious feeling give practical information about the mutual orientation and movement speed of the body modules [35–38].

Type I mechanoreceptors are usually composed of 3–8 recording structures (40–100 μm in size) and are usually located in the outer (fibrous) layer of the joint capsule. Their characteristic feature is low excitability threshold, enabling the generation of positional information about the angular position in the joint.

Type II mechanoreceptors are usually built of 1–2 recording structures (100–280 μm in size), located in the inner layer of the joint capsule. Their characteristic feature is low excitability threshold, as well as generating information about the direction of motion.

Type III mechanoreceptors are composed of 1–2 registering structures (100–600 μm in size) with high excitability threshold. Being located in enthesis, they generate alarm information about muscle and tendon overtension.

Cutaneous mechanoreceptors respond directly to the movement (as a change in shape) or indirectly, through skin contact with clothing and other objects. Type I cutaneous mechanoreceptors, known as the Meissner bodies (detectors of motion speed) and Merkel bodies, are most commonly located on the palms, feet, and lips. They are characterized by a lack of spontaneous activity and high sensitivity to skin movements, allowing for surface type and shape cognition. Type II cutaneous mechanoreceptors, known as Ruffini and Pacini bodies, are most often located on palms, feet, and trunk. They display spontaneous activity for skin deformations and stretching resulting from mechanical stimuli in areas distant from the receptor. In their action, they have a static component, dependent on the strength of the stimulus, and a dynamic component, dependent on the rate of parameter change of the stimulus [39]. The cerebellum receives and triggers information from the vestibular receptors, proprioreceptors, exteroceptors, and telereceptors in the functional buffer, forming model equivalents of the state of balance at T1, T2, etc., Tn, causing the evolution in the center of gravity's motion toward the central area of the supporting plane's critical curve. When the pyramidal tract is activated by the control sequence of conscious movement, especially when there is a significant shift of the center of gravity, it activates an involuntary, multi-muscle sequence of movement, resulting in balance correction [27].

4. Model and outline of balance mechanisms

The simplest model describing standing posture is the inverted pendulum model. According to it, stability in standing position requires data regarding the position of the upper end of the pendulum (head), as well as monitoring the rake angle relative to the supporting plane. The main mechanism is the proprioception of pressure distribution of the ankle joint surfaces, changes in lower leg muscle's length and tension, as well as angle changes between axis of the foot and lower leg [40].

Balance loss prevention is effective provided that the nervous system is able to recognize in less than 70–100 ms, the characteristics (mainly direction and force) of a destabilizing stimulus, and raise a competent engram containing a set of drivers for adequate muscle synergy (rapid coordinated movements compensating instability), which would restore synergy [41]. The speed of the motor reaction (which restores the balance) to the stimulus decreases in proportion to the number of alternative patterns of motion behavior, existing in subject's memory. Hence, the adoption of a specific position (bending body forward) limits the choice of a large number of possible alternative movement patterns, reducing time of a proper coordination scheme to access the motoneurons of the pyramidal and extrapyramidal tracts [42]. It reduces the time to generate a motion sequence, which corrects the displacement of the center of gravity outside the critical curve of the supporting plane, increasing the stability. This behavior can be seen in young people, walking on an unstable surface (adaptation), or in the elderly, even on a stable surface (involution). A strategy for (slightly disturbed) balance recovery in subjects standing on a stable surface has been described, where corrective sequence begins with the contraction of the ankle joint muscles (ankle joint strategy), as well as another strategy, in people standing on a narrow ground, which begins with thigh and trunk muscle contractions, and further includes lower limb muscles (hip joint strategy) (Horak, Neshner). The third way for balance recoFect the body from falling down (step strategy) (Nesher) [20].

5. The confrontation of models with the real neuromuscular system of insects

The anatomy of the insect's locomotor system includes all elements described by biomechanics, i.e., limbs, whose rigid elements are combined in a flexible and repeatable manner, with a certain number of degrees of freedom and muscles, with a strictly defined topography of attachments. This design makes it possible to create simple and clear vector diagrams of internal forces, as well as easy, quadroscopic marking and registration of insect anatomical parts, in particular limbs. Compared to vertebrates, the structure of the nervous system in insects is also easy to read, ranging from the neuromuscular junctions, sectoral coordination elements, to the central nervous system. These features create a good chance of learning the full standard species topography of neuron and neural tract clusters and, with its help, determining the list of simple and complex reflexes and understanding their functional foundations. The listed anatomical advantages and high availability of these animals make them particularly useful in the study of parametric motor-behavioral standards, which are the starting point for the next stage, namely, modeling of elementary and complex reflexes, or holistic models in the information space, e.g., neural networks. Particular attention should be paid to basic research regarding:

1. Maintaining balance in the standard position while performing individual movements and walking

2. Creating basic, complex, and systemic models of motion control [43]

3. Reasons for making a movement [44, 45]

4. Modeling of movement in a simple network of an insect's ladder-like nervous system [46]

5. Spatial orientation [46, 47], based on optical (sun, moon) [48], acoustic [49], and magnetic markers [25, 50]

6. Standards of insect motility: an area of inspiration for robotics

The movement coordination schemes obtained in animal studies are an inspiration for the construction of control algorithms in robotics [51, 52] and prosthetics, which supplements the functions of lost limbs [53, 54]. Equipped with receptors, a simple nervous system of an insect probably allows to define the internal state of the body against the background of a model composed of environmental parameters registered by telereceptors, using projection, symbolic coding, and actions on models, which are an electro-resonant representation of the real phenomena [55–57]. Is it possible, then, to integrate spatial multi-receptor models in simple insect neural networks [58, 59]?

7. Biomechanics in the chain of driving phenomena and movement consequences

The ability to maintain balance and the ability to navigate in the outdoor based on specific markers, including the sun, have been well established in simple nervous systems of insects [60]. However, referring to the earlier discussion, human physical activity covers a broad set of behaviors related not only to maintaining balance but also to the locomotion specificity, facial expressions, speech, and voluntary movements of varying complexity. A stable posture is a prerequisite for the majority of voluntary movements, locomotors, and creative activities. The describable human body motion should be treated as a chain of mutually coupled procedures: perceptive, decision-making, control, motor, and systemically interactive and correlated with internal and external reference system. The biomechanical (dynamic) description is a set of parametric sequences of time-synchronized shift vectors of specific marker points [61, 62]. Movement anatomy should be recorded and evaluated in full anthropomotorical context, at least in four categories:

1. Biomechanical, taking into account the range of joint motion, the balance of motion vectors, the muscle strength, and the muscle tone

2. Coordination, understood as the evaluation of the speed and precision of the observed movement patterns

3. Sensory, regarding the sense of the body setting in space and direction of movement, as well as body orientation in relation to the gravity arrow

4. Psychomotor, analyzing the impact of emotional state and psychopathology on the parameters of vector motion analysis

Practical attempts undertaken by the authors in this regard have been successful [63–65].

Adopting an upright position is a reflex action, based on evolutionary and ontogenetically reproduced postural reflexes. Evolution has created a universal movement calibration system using a gravity arrow available all over the planet. The resulting habit of referring motion vectors to a universal direction (even in the darkness) is one of the last stages of mastering and improving motor skills, which allow for a multiple repetition of motor task in a similar manner without involving the consciousness. In children aged 7–11 years, there is an escalation in the development of balance capacities, which is probably never repeated again. The manifestation of unconditional reflexes in this period induces the child to experiment with motion and consequently leads to the formation of individual conditional reflexes, then dynamic stereotypes, and eventually motor habits. Due to the fact that a little later, between 11 and 13 years, there is a period of temporary stagnation or even partial regression of internalized behaviors, it is very important for schoolchildren to have a well-thought-out training, focused on developing coordination and balance skills.

The main area of research in contemporary anthropomotorics is the search for a motion control model and defining methods of encoding information in these models [6, 66, 67]. Drawing inspiration from the mathematical description of the methods for rigid bodies, it is worth remembering the significant differences between the machine and a living organism. A machine, described globally or at subsequent stages of the sequence of its components, performs a specific move (always in the same manner), precisely predetermined by the design plan as well as by the control variant chosen. These features are clearly defined by the machine's working element in the number of degrees of freedom, and the applied constraints accurately determine the specificity of the working track, which will be implemented (with the same parameters) for any number of repetitions. Living organisms can achieve a similar kinematic goal, for which they use an interactive sensorimotor procedure; however, in view of the necessity for a greater number of repetitions, they carry out experiments in the field of kinematic path control (from state A to state B). These experiments consist in attempts to reduce degrees of freedom in the movement control procedure, which also means the reduction in the number of muscles involved and energy saving in consequence (Bernstein). Improving the kinematic forms of movement while reducing energy consumption leads to optimization in the species development of living organisms [68].

It is worth adding that after a thorough, cytoarchitectonic definition of a large part of the cortical sensory fields, which have a strictly defined anatomical relationship with the ascending spinal cord pathways and receptors, as well as the cortical motor fields associated with the descending spinal cord pathways and muscle effectors, a research for functional relationships of higher order between different parts of the cortex began [23].

The above processes for the reduction of degrees of freedom, which aimed to optimize human and animal movements, are described by Bernstein in his theory of control levels. This theory seeks to outline and highlight within the central nervous system (CNS) a specific number of hierarchically associated classes of functional creations, responsible for movement creation of certain specificity, which significantly decreases probability of a control error [69]. Gradually, a cause-and-effect relationship of four factors that determine mobility began to emerge:

a. Excitability (ability to receive and respond to environmental stimuli)

b. Analysis of the environment and creationism (perception of new motor tasks and the need to solve them, which leads to the development of existing or emergence of new executive organs, as well as emergence of new formations in the CNS, enabling the creation of new sensorimotor abilities)

c. Optimization (development of versatile sensorimotor capabilities that enable building new skills in terms of complex motor tasks)

d. Sensorimotor intelligence (cleverness, artfulness, and agility combined with prospective calculation of motor task in the context of the structures and entities recognized in the foreground and background of the subject's physical environment) [22, 70]

Initially, a good reflection of it was Bernstein's model, consisting of five hierarchically coupled levels of motion control, trying to find constituent components in complex movements: (A) muscle tension, (B) muscle synergy, (C) spatial field, (D) complex activities, and (E) symbolic operations. Such an arrangement partly explains the follow-up functions of the extrapyramidal system, which adjusts the geometry of angular parameters of the body's rigid modules (bones-joints) and the degree of the bond stiffness (joints) and viscoelastic systems (muscles), in order to maintain balance in the gravitational field, regardless the temporary destabilization of one of the modules of the body (e.g., hand) performing a conscious movement [71]. According to this theory, the system of levels is configured hierarchically, from superficial to profound layers. This direction is caused by the integrating function of layers D to E, which stimulate performance of the motor task components without involving conscious attention. For these levels of movement structure, there have been introduced various control models, for example, for (B), Gibson's ecological model; for (C), the hypothesis of an equilibrium point; and for (D), Schmidt's cybernetic model. Moreover, for reality mapping and motion control, levels A, B, and C use sensory code, while levels D and E use symbolic code of motion representations [20].

The introduction of electronic inventions into biomechanics caused an avalanche of new discoveries in the field of motion kinematics, and it became necessary to modify Bernstein's theory, where instead of a static, layered system of control levels, a dynamic model has been introduced, whose subsequent layers overlap to form transition zones. This new feature well explains the fact that the development of individual levels proceeds gradually, beginning with the emergence of new perceptive abilities and only then of new motor skills. It was assumed that the new level of perception may to some extent define the demand for cooperation with the structures having simpler coordination engrams, which may be a factor that powers the development of sensorimotor and control abilities. In addition, it was assumed that, stimulated with need, a lower level of motor behavior organization, having used up all available range of options, may command the higher level to seek for more optimal motor compositions, as it possesses greater associative capabilities. In this model, there is thus a two-way flow of directives regarding functioning, namely, from the top to the bottom (in order to include simple coordination engrams in the complex, three-dimensional diagrams) or from the bottom to the top (in order to adapt existing, underlying motion engrams to the new kinematic situation in a given environment).

In the complex procedure of balance creation, the cerebellum receives and triggers information from the labyrinth receptors, proprioreceptors, extroceptors, and telereceptors in the functional buffer, thus forming multi-loop models, which are electric-resonant equilibrium equivalent at T1, T2, etc., Tn. These models are confronted with many equilibrium patterns; they create corrective engrams that cause the evolution of motion of the body center of gravity to the central area of the critical curve of the supporting plane. When the pyramidal paths are activated by the control sequence of conscious movement, especially when there is a significant shift of the center of gravity, it activates the independent, multi-muscular movement sequence, causing balance correction [72].

7.1 Perception and gnosia

A perceptive-gnostic procedure occurs in the human CNS (and probably in many other vertebrates), and despite the fact it includes projection engrams, simple and complex perceptual engrams, gnostic engrams, model intentional engrams, and decision-executive engrams, it is not available for measurement systems of classical biomechanics. In this procedure, in a specified period of time, stimuli are projected from proprioceptive receptors (strain gauges, located in tendons, muscles, and ligaments), extroceptors (skin sensation), and telereceptors (balance organs—gyroscopic accelerometers, retina, eye receptors). In areas of the sensory cortex, anatomically associated with the appropriate type of receptors, electric-resonance, loop equivalents of phenomena recorded by the receptors are created, namely, projection proprioceptive simulation, projection extroceptor simulation, projection telereceptor-gravitometric simulation, and projection telereceptor-visual simulation simulator. During this time, cognitive processes also occur, referring the resulting projective equivalents to the appropriate pattern bases, resulting in the creation of gnosia, or engrams with the meaning defined for the subject.

7.1.1 Resonance functional integration model

Illustratively, one could call these cerebral, loop-electro-resonance equivalents of phenomena recorded by receptors, for example, proprioceptor-cognitive simulation, exteroceptor-cognitive simulation, gravitational-cognitive simulation, visual-cognitive simulation, and auditory-cognitive simulation. The three-dimensional association on the common timeline of local, organ-specific electro-resonance models of receptor perception and gnosia results in the creation of a conscious, multi-level conjugated cognitive simulation, which is a representation of a body model integrated into the model of the environment, with particular emphasis on sensory receptor density zones within the eyes, mouth, tongue, and fingers. Corrected by gravitational and geometric markers of the environment, model equivalents of sensory phenomena in the cortex of the brain give the possibility of reflex orientation of the head and long axis of the trunk and extremities in relation to the gravity arrow. The first step to gaining the awareness of being is the perception of the model presented above, determining the geometrical features of the shape of one's own body and the environment, as well as proportional relations between the elements of the environment and the body. Therefore, the model gives a sense of the shape and integrity of the subject's body functions in relation to the three main space vectors and components of the internal environment.

It is worth adding that the conditioning of the emergence of consciousness, the subtle state of simultaneous, inter-center synchronization for the cortical projection and gnostic structures of the brain, requires an independent timing system that determines the excitation (specific resonance frequency) of the simultaneous activity of specific brain areas, constituting the electron resonance generating neural medium in a given time-interval loop of currently interfering components of a conscious being. In the model proposed by the authors, the timing system, using a specific, unique frequency, synchronizes interference groups of neurons to a synchronous electrical activity. A spatial, three-dimensional, subtle interference space is created, composed of variable electric fields of neurons, for which the criterion of integrity in relation to other (surrounding) nerve cells is a specific resonance frequency. An important criterion for the functioning of consciousness, in terms of the clock frequency of the timing system, seems to be the limiting frequency conditioning the overlap of the descending edge of the dying pulse on the leading edge of the next pulse. In the proposed model, there is a resonant coupling with a set of cortical, projection-gnostic representations of specific receptors or decoupling, which allows the transfer of conscious attention from one area of receptor gnosia to another.

The association of internal reference system parameters with external reference system parameters in a synchronous manner with the internal awareness timing system makes it possible to define parameters for the last in the sequence of a motion event—the endpoint controllability [71]. It seems that the registration of the end position signal for the motor coordination engram (which currently has exclusive access to the neural space of the pyramidal pathways) releases him from the position of a leader and activates the mechanism of changing the engram, by way of competition or probabilistic autocreation [73].

7.2 Decision-making procedure

As already mentioned, an important element of the consciousness model is the qualitative recognition of the specific features of the model projection of receptors, based on their relation to the relational database in the memory of the subject and after obtaining the conjunction—recognized as specific structures or phenomena. Perception allows to create geometric relationships between one's own body dimensions and the elements of the environment, as well as emotional relationships in relation to one's own body and its relationships with other subjects and the surrounding structures [74]. The decision-making procedure is based, inter alia, on the recognition and classification of surrounding objects in terms of the possibility of their use for the implementation of a specific task. The predefined gnostic parameters determine the structure and range of operation of the seeking system in order to isolate an adequate executive engram, characterized by a stable structure [75].

This involves a whole range of applicable motor strategies that aim to solve specific motor tasks. The brain has or, in a probabilistic dimension, creates a number of alternative strategies for sequential-functional control of muscle groups, which are concepts for solving a planned vision of a kinematic situation [6]. They are created as intentional models of movement that compete for access to the area in control of neuromuscular system in the motor cortex. The neural network makes a directional selection of models, using (as criteria) parametric data acquired from both exterior and interior of the body, choosing the most optimal algorithm in a given kinematic situation [71, 76]. Once given the priority, the selected model becomes an engram of motion creation and takes over the control of the motor cortex, thereby yielding a functional access to the pyramidal tract [77–79].

The control procedure, which is an engram containing a set of time-oriented motoneuron control procedures, is sent efferently through many parallel pyramidal pathways, stimulating the muscles to a conscious, coordinated movement [80, 81]. Motor procedure—the stimulated part of the muscular system—derives energy from the stored high-energy phosphates only in the first seconds of contraction, and further maintenance of physical activity depends on the efficiency of stimulating the Krebs cycle oxidative reactions, as well as efficiency in the removal of the waste products. The above example gives an idea about the interference depth of kinematic stimuli in the body's molecular phenomena. The interactive procedure systemically affects the biomechanical consequences of intentional movement that could threaten the balance of the whole body. This procedure forces involuntary, follow-up reflexes from the extrapyramidal system, which modulate skeletal muscle tone in different parts of the body, allowing to maintain the balance and constant direction in relation to the gravity vector. A living organism, in contrast to a comparable (in terms of size) and stable (in given circumstances) crystal structure, is a visco-elastic body that changes its characteristics locally, rather than in a way that would be optimally adapted to the forces acting from the outside. Every kinematic situation, which can be described using a simple marker geometry, has a qualitatively different way of linking active information conveyors (nerves), energy converters (muscles), and the blood distribution, which conditions the efficiency of locally intensified metabolism. Local blood distribution profile is connected with ion concentration changes, which by means of the changing tissue resistance, impedance, and magnetic induction influence the sensitivity of the skin sensors. These parameters can be registered using electrodiagnostic methods, such as ECG, EEG, EMG, ENG, and EEA. It can therefore be assumed that each of the organ pairs existing in a living body, despite their high degree of structural and functional autonomy, functions in an information-coupled way through mirror elements of the nervous system and also in a hormone-distribution-coupled way, realized through the vascular system. In order to obtain a complete dysfunction image, it is therefore not sufficient to merely evidence asymmetries in the goniometric tests, as neurogenic, vascular, and immunological causative factors should also be taken into account.

7.3 Control strategy

One of the control strategies aimed at maintaining the general symmetry plan within a pair of organs is their individual endeavor to develop their own existence, which manifests itself in the competition for access to the distribution of nutrients and a tendency for constant reconstruction and redevelopment. In case of damage in one part of an organ pair in the nervous system, at least two basic survival strategies are launched, the first of which consists in intensification of the vascular perfusion and activation of neuronal paracrine secretion, conducive to hypertrophy and hyperplasia of cells in the damaged organ. The second strategy consists in increasing the metabolic efficiency of an intact organ, increasing its contribution to maintaining steadiness of parametric balance at the central level. Organ reflexes for the first strategy are located at the level of the lowest spinal cord integrators and for the second at the level of high spinal cord integrators and hypothalamus [82, 83].

The algorithms shaping motor and metabolic coordination in a living organism are interactive, held under the form of the convergent or opposing oscillations around certain equilibrium points. The extent of this oscillation, being the body's response to the changing environmental stimuli, depends on regulatory efficiency on kinematic, computer, and metabolic level of the system. A young organism, not having many coordination engrams, reacts cautiously, using a large number of oscillations with high parametric amplitudes, trying to optimize to the equilibrium points. A mature organism, having a shaped profile of movement strategies, adapts its behavior to the situation a lot faster and moreover in a balanced and elegant manner.

Cellular homeostasis manifests itself by an individual cell's ability to develop functional states with parameters that pose no threat to the system itself. Its higher level—the tissue homeostasis—is a feedback of many metabolic oscillators, whose plasticity and tendency to minimize the number of energy transformations ensure consistency of the structure. The main tasks in this regard include the temperature and ion composition stabilizing, the regeneration of the stroma, data mediums and enzyme systems. In emergency states, the majority of chemical reactions in a living cell can occur at some distance from the optimum; however, the thermodynamic efficiency of a given transformation decreases in such case. The amount of energy obtained per a substrate mass unit diminishes, with a significant increase in the production of waste products. On the organ level, such a situation is identified with the developing disease process, in which a large role is played by the control disorders.

A prolonged dysfunction affects the structure of the individual cells or their groups, leading to the reduction in their functional reserves and slowing their recovery, with a steady increase in the rate of the waste substance accumulation. The conjunction of these processes causes the system to reach the control point, trespassing which starts an irreversible process of the apoptosis. Persistent regulatory metabolic disorders can lead to permanent organic changes. Information from skin, muscle, tendon, and visceral receptors regarding the influence of the environment on the body, through the ascending paths, reaches the central nervous system, where it is recorded and analyzed. Depending on the final response to the stimulus at the level of the core, hypothalamus, or cortex, the body reacts with an autonomic reflex or conscious action. In the case of conscious actions, it can be a deliberate motor reaction, while an example of a reflex follow-up reaction is the accompanying movement of the tension of other skeletal muscles, ensuring the body's balance. Autonomic reflexes occur without the participation of consciousness, enabling, for example, automatic glycemic control, symmetry of blood distribution in micro- and macrocirculation, or regulation of endocrine gland secretion. This often causes the subtle molecular response to environmental stimuli.

8. Modulation of the body motor parameters

There are many factors affecting the precision of the targeted and follow-up movements such as fatigue [84]; emotional conditions [85]; mental disorders, such as depression and schizophrenia [86]; postictal conditions [87]; extreme sports [88, 89]; and accidents.

Pain modifies the body movements in a variety of ways. Being an alarm reaction, in the first place, it leads to the reduction in the range of motion along pain-causing, collision trajectories, compensating them by the development of nonphysiological or paraphysiological paths, known as the asymmetric profile of dysfunction. Unilateral reduction in the amplitude of motion, along with compensatory increment in another, usually symmetrical area of the body, disturbs smooth, alternating symmetry of muscle work. Virtually every clinical problem that is directly or indirectly associated with the gait function affects the symmetry of foot, knee, and spine load and consequently the degree of their functional and adaptive reserve utilization [90].

The consolidation of the asymmetrical compensatory movement pattern leads over time to more or less advanced anatomical modifications, including muscle, joint, tendon, and ligament remodeling and even remodeling of the bone structure. These changes are noticeable not only due to the use of advanced imaging techniques, such as CT, MRI, and PET, but also on classic radiophotography images. The problem is only that in the medical community, because attention is still not being paid to looking for the causes of apparent asymmetries. A good example would be an X-ray of both femurs, one of which has a clearly thinner cortical layer. Radiological descriptions rarely suggest that one of the patient's legs is shorter and that most of the body weight (>50%) is transferred to it, which results in remodeling and thickening of the cortical bone layer [91]. The evolutionally accumulating experiences of species that, apart from the rigid behavior of the hypothalamus in the cerebrum created the seeds of abstract thinking, tend to quickly expand the strategy of survival in the environment by joining simple unconditional reflexes—complex sequences of behavior aimed at classifying the environment into neutral, health-promoting areas (vegetation, water reservoirs, mineral deposits) and clearly hazardous (e.g., deposits of toxic substances).

8.1 Elements of the movement metrology

The symmetry of the structure and function of the basic elements of the human body remains in a close cause-and-effect relationship with the symmetry of the structure of the nervous system, treated as a control system, as well as the vascular system, which is in fact a supply and control system for the cellular stroma. Therefore, it can be assumed that each of the organ pairs existing in the living organism, despite the high level of structural separation, functions in an information-coupled manner by means of mirror elements of the nervous system, as well as in hormonal-distribution couplings implemented through the vascular system. The balance of the neural network conducted in subsequent integrators of the spinal cord and hypothalamus, in terms of controlling the distribution of signal and nutrients, including the mechanical and energetic loading of pairs of twin organs, allows their maintenance in a state of functional and anatomical symmetry.

The macroscopic effects of maintained symmetries (especially in the geometrical range) currently belong to the basic criteria for assessing the condition of the musculoskeletal system. However, using the bioengineering assessment of the disease, which occurs with one limb dysfunction, it can be presented as asymmetries of individual parameters or parametric sets, not only regarding the location of markers of body movement in 3D space but also muscle strength, blood supply, resistance, temperature, and skin sensitivity to stimuli, while the numerical determinant of these asymmetries decreases from unity to 0, as the dysfunction increases in the course of the disease and increases again (to values close to unity), in a manner proportional to the disappearance of the symptoms of the disease [92, 93].

In this context, it is worth paying attention to the need for a reliable, parametric and quantitative estimation (that corresponds with the degree of biomechanical dysfunction of the musculoskeletal system) of a greater number of internal body parameters, whose mutual relationships shape the causative dimension of the noticeable biomechanical asymmetries, as well as the disease-related feelings of the patient. It is worth to note that it is difficult to talk about a proper estimation of the current state, prognosis, and treatment effects, without a reliable estimation of the quantitative parameter. Therefore, methods for objective monitoring of disease parameters are actively searched for. Great importance is attached to diagnostic imaging, pathology, and electrodiagnostics. The criterion systems existing in modern medicine give high sensitivity and specificity of qualitative diagnoses, in particular for diseases with a well-recognized etiology and mechanisms of action. An experienced doctor examining a patient is able to see in his body many deviations, such as compensatory movement profile asymmetries, static kinetic reflex disturbances in the Romberg test, temperature and humidity asymmetry of limbs, difference in skin sensitivity to pain and touch, and asymmetry of limb mobility and reflexes; however, he has a difficulty in parametric and numerical estimation of the characteristics of his observations. Estimated methods still exist in the practically used scope of locomotor system assessment, based on interactive, subjective relations between the patient and the doctor, such as the older Lovett scale, visual analog scale (VAS), or newer quality of life scales [94, 95].

9. Medical examination

The metrological dimension of daily medical examinations of the musculoskeletal system is usually limited to the use of simple measurement techniques and tools, due to the difficult access to expensive, advanced measuring technology. In addition, although there are many sites with advanced quadroscopic and strain gauges for assessing and recording biomechanical parameters, their presentation and data collection systems are usually incompatible on an inter-center scale.

The first element of this procedure, an interview, organizes knowledge on the subjective symptoms of the patient. The study design is layered, because it begins with a general health assessment, and delves into the area of details that are specific for certain organs, so it is relatively easy to reduce it to a survey test, allowing for an approximate quantitative assessment. The quality of life rating scales, anatomical-functional syndrome assessment scales, and pain assessment scales have gained in importance here. The credibility of this information depends on:

1. The level of intelligence and education of the patient

2. The specifics of a physician's contact with the patient

3. The general condition of the patient, in particular the duration and severity of the disease

4. The level of intelligence of the patient, which determines the clarity of the message

5. Patient's current attitude, modifying expression towards aggravulation or dissimulation

6. Existence of nonmedical intention (desire to obtain compensation or disability benefits)

The doctor's education, clinical internship, and experience as well as his condition on the day of the examination determine whether the collected information will be used as an inspiration for a series of diagnostic associations.

The physical examination is an assessment of the patient's state of health with the use of the telereceptors (eyesight, hearing, smell) and extroceptors (touch, heat, cold, movement) in palpation. The next stage of the study is the use of minimally invasive stimulations in the form of tactile and tactile stimuli, provocative tests, or simple measurement tools. Subsequently, tests are carried out using a goniometer, plurimeter, plumb line, and linear bearing, which allow simple numerical approximations, unfortunately often unrepeatable, due to the fact that every physician introduces some alterations (based on his/her own experience) into the measurement standards. In practice, it translates into introducing one's own interpretations or minor methodological mistakes, specific for one's habits referring to measurement position and the method of the measurement tool's application. These phenomena, provided that patients are examined by the same physician, generate a uniform and distinctive profile of system errors that are fairly easy to eliminate, but when these patients deal with other doctors, they are confronted with a different interpretation profile of the measurement principles (system errors). The overlapping of different approaches to accomplish the same measurement task often leads to the interpretation differences, not only in terms of quantitative assessments but even of the qualitative criteria [61].

10. Metrological engineering techniques

Taking into account the above issues, the pursuit of professionalization in the measurement process gains extreme importance, which can be seen in the works of many authors, concerning the static features of the spine using the projection moire, axial spondylometry, as well as changes of these characteristics in various clinical conditions. Much attention was paid to the assessment of the motion range (SFTR), balance, and gait analysis carried out using contact and photogramometric methods. Joints are an area of particular interest—in clinical practice they are most often subjected to endoprosthetic surgery, i.e., knee, hip, shoulder joints, intervertebral discs or other elements of vertebroplasty, and even the joints of the fingers. An interesting combined research began, concerning the synergy of dynamometric tests, the range of foot movement and strength of lower limb muscles, and gnatosomal analyses. There are also attempts to model complex limb functions based on motion measurement data and EMG signals. Measurements using bioengineering techniques have a great impact on the development of clinical metrology. They have many advantages, such as (1) standardization of the measurement conditions (measurement track in particular), (2) repeatability of the body stimulation scaling parameters using electronic systems (3) repeatability of scaling of the measurement parameters, and (4) stability of the calculation and result-interpreting criteria. Noninvasive recording techniques, consist in resting or functional, but contactless monitoring of vital functions, based on registering parameters that are spontaneously emitted by the body, such as infrared radiation (pyrometry, thermography), electric field (electrography), or magnetic field (magnetography).

Contact recording techniques consist in applying a sensor to the skin surface, in order to establish physical contact (often galvanic as well). The skin surface changes a number of its properties depending on the symmetry and specificity of the processes taking place in the internal organs; hence, the registration of parameters at specific points of the marker will give clear information about the phenomena occurring in the internal organs (e.g., thermometry, ECG, EEG, EMG). The essence of measurement techniques is diverse, depending on the degree of invasiveness [93]:

1. *Bilaterally interactive test*, in which the doctor affects the patient's body with a mechanical stimulus that is hard to repeat (e.g., pricking the skin with a probe or strike) and then the patient makes a subjective response assessment, based on which the diagnostic conclusions are being drawn.

2. *Unilaterally interactive test*, in which the doctor is replaced by an electronic device that generates parameterized and repetitive stimuli while the patient subjectively determines their perception threshold (e.g., palestesiometry, audiometry).

3. *Parametric telemetry test*, in which parametric responses of the patient are recorded—they are based solely on the parameters spontaneously issued by the body (thermography, electrography, magnetography).

4. *Parametric contact test*, involving, for example, fixing a sensor on the body (or unipolar group of sensors), which provides information on selected parameters of the body.

5. *Bilaterally parametric test*, where the body is influenced by a reproducible, parametric stimulus generated by the technical device (which can either

remain in physical contact or else operate contactlessly) while the evaluation takes place outside the sphere of the patient's consciousness, through the system of sensors registering involuntary vegetative reaction parameters.

6. *Invasive registering techniques (A)* that consist in making (in relation to the examined body) energetic extortion as well as recording and interpreting effects of this extortion in the form of (a) rays reflected off the surface (photography), (b) rays penetrating the object (X-ray, CT images), (c) the object being penetrated by the magnetic field (NMR), and (d) the object being penetrated by the stream of ultrasounds (USG).

7. *Invasive registering techniques (B)* that consist in introducing into the body complex chemical compounds, which on one hand have affinity to specific body structures and, on the other, contain a substance that emits radiation that penetrates the coating (usually gamma)—this emission is recorded by the external recording system (scintigraphy, gammagraphy, single-photon emission computed tomography (SPECT)). Based on the above emission, a spatial reconstruction of the organ image is prepared (based on areas with isotope concentration).

11. EBM

A huge problem that is still present in movement studies and physiotherapies is the tendency to look at the multidimensional profile of the patient's suffering and dysfunction through the prism of a narrow specialization, causing multithreading and lack of synergy in procedures related to the diagnosis and treatment of a complex locomotor dysfunction. In this context, a complex biomechanical, psychophysical, and clinical problem begins to be seen as the sum of some separate sequences, which are diagnosed and treated fragmentarily in the narrow ranges of specializations of the successive therapists.

It leads to an absurd situation, in which a pain and dysfunction problem is being taken care of by subsequent specialists in physio- and kinesiotherapy and finally in biomechanics. Based on their own experience and simple measurements, they assess the patient's condition and plan a simple treatment program, mainly in the aspect of symptoms and function, not having a full insight into the interpretation of the results of advanced diagnostic tests. Diagnostics, in turn, is carried out by a team of doctors, who have direct contact neither with a variety of procedures in applied physiotherapy nor with their effects. As for issues revolving around the experience of pain, especially against the background of the patient's individual psychophysical characteristics and predispositions, they are estimated by the psychologist, who is deprived of a broader knowledge of the details of diagnosis and therapy specifics.

In this context, there was a need for a systemic change in quality, and thus also the effectiveness and image of modern physiotherapy, beginning from function in analogy to the principles of EBM presented above, as evidence-based physiotherapy (EBP). The trend initiated a few years ago was reflected in the literature, in the form of interesting publications, displaying various aspects of this issue. One of the first, conceptually well-structured works was a manual [96], then a collective work [97], which postulate that a physiotherapist performing direct, manual diagnostic and rehabilitation activities in a patient, should use the results of advanced clinical diagnostics to clarify any doubts arising from direct observation of the patient. In 2001, the World Confederation of Physical Therapy (WCPT) Expert Meeting on

Evidence Based Practice in London 2001 was founded in London. An important initiative here was to create a PEDro system database accessible via the Internet.

In this new approach, EBP is a collection of clinical procedures that aim at the best possible, holistic use of the therapist's clinical experiences, supported by (verified and available in practice) scientific evidences of image, laboratory, and functional and psychological diagnosis, whose application and integration in one personalized clinical picture increase the effectiveness of medical procedures and patient's safety in the process of the musculoskeletal system's diagnosis, shaping the highest possible level of physiotherapeutic treatment [97]. It can be noticed that the presented definition is quite convergent with the Polish model of the rehabilitation team, promoted since the 1970s (Milanowska, Dega), which uses the consular way of disciplinary knowledge of a rehabilitation doctor, physiotherapist, psychologist, and social worker in order to find a comprehensive solution to the problem of motor organ dysfunction in a medical, psychological, rehabilitation, and social aspect. The cited textbook is particularly recommendable due to the fact that the authors have made a very comprehensive summary of the existing literature on the subject, outlining a very clear picture of the issue. Thus, EBP is shaping a directional model for the development of new science of movement disorders, which are the essence of many medical and physiotherapy specialties, aiming at replacing the currently dominant profile of the assessment of movement disorders implemented by means of distributed sequences, narrowly specialized engineering research, and general medical procedures for the creation of multidimensional and a personalized image reflecting the tangle of causative problems of pain with parametric disorders of biomechanics of movement, combined with metrology of psychological and social consequences.

It is therefore of the utmost importance that the professional evolution of specialists dealing with the musculoskeletal system does not stop at specialized, narrow clinical, or metrological competences but aims at increasing interdisciplinary competences, which will result in the ability to perceive parametric body movement disorders against the background of causative phenomena and above all their adaptive and clinical consequences. In such a new situation, a physiotherapist with broadly formatted practical and theoretical knowledge will not only receive but will also be able to use access to team knowledge, including clinical information on the etiology of the disease, as well as to the results of detailed imaging and clinometric tests. The planned treatment strategy will gain a new, more pragmatic parametric foundation, which in addition to diagnosis offers the possibility of metrological supervision of the effects of therapy, including psychological clinimetry. It is important to remember about the changes in the patient's psyche who is suffering from pain, as well as limiting the implementation of his/her priorities on personal, professional, and social grounds. It is also very important that the technical specialist who formulates the equations of movement does not limit his interest only to the movement of markers in three-dimensional space or tribological relations on the rolling surfaces of the joint, but be aware of the causative phenomena of movement, as well as various consequences for specific parametric intervals. Extending technical knowledge with a biological dimension will certainly result in the improvement of existing and the use of new mathematical applications.

12. Quantitative assessment of the observation

The ability to make an effective, quality diagnosis of the patient is a kind of subjective and highly individualized diagnostician's skill, which combines the elements of systemic, scientific knowledge with acquired experience and intuition into one coherent system. During the professional development period (of a physician, for instance), this trait evolves as a result of one's dominant character, the amount of acquired theoretical knowledge, and the number of known clinical cases. A unique element—human nature—is therefore a modulator in the way of perceiving subsequent clinical situations, in particular seeing them through the prism of acquired scientific knowledge and gradually collected capital of experience.

Many years clinical experience has been reflected in the creation of more and more specific criterial systems, being a scientific generalization of practical observations of many generations of doctors. These systems provide high sensitivity and specificity of diagnosis in terms of quality. At the moment it can be said that for the diseases with well-known etiologies and mechanisms, criterial systems do serve their purpose. An experienced doctor, while examining a patient, easily sees the qualitative characteristics of compensative asymmetries in patient's profile of motor dysfunction. For example, he/she notices asymmetry and direction of compensation in statokinetic reflexes in the Romberg test, draws attention to the asymmetry in the SFTR measurements, differences in temperature and humidity of the limbs, differences of the skin sensitivity to pain and touch, and asymmetry of myotatic reflexes, but **has difficulties in estimating the quantitative characteristics of his/her observations**.

To sum up, one of the major problems of musculoskeletal medicine is the difficulty of a reliable, quantitative estimation of biomechanical parameters of the asymmetry of the patient's adaptive dysfunction profile and the subjective dimension of his suffering that occurs in the course of a qualitatively diagnosed disease. It should be remembered that without a reliable assessment of the patient's initial biomechanical parameters, it is difficult to talk about proper assessment of treatment effects and prognoses. Currently, in the evaluation of the musculoskeletal system, estimated methods are still in use—they are based on mutually subjective relationship between the patient and the physician, with examples such as the Lovett and WAS scales or (more recent) quality of life scales. Metrology-wise, the use of simple measuring tools dominates and is rarely supplemented with individual cases of only locally available (and usually unique), advanced measurement technology.

13. Clinimetry

Methods of converting specific (but descriptive) clinical criteria (that are specific for different disease units) into the form of bioengineering parameters, which would finally allow for a quantitative monitoring of disease parameters, are being actively searched for. Another question is whether a principle that is obvious for the angular, vector, and time parameters used in biomechanics can be applied in the description of results of medical imaging, pathology, and electrodiagnostics. There is a lack of a supervisory system that would function as a standard, inter-center way of presenting well-known measurement technologies enriched with a quantitative aspect adjusted to this standard, designed for communication and capable of assimilating new diagnostic solutions to suit the presentation and systemic work.

13.1 Movement marking

An important element of bioengineering research is the correct location of the body marker points, thereafter subjected to a static 3D parameterization, with the possibility of tracking, associated with the movement of the parametric drift in the dynamic graphic analysis or in telemetric analysis of radio markers. A characteristic feature of these points must be location stability as a common feature of individuals that belong to the same species. Thus, a body construction property, acting as a geometrically distinctive feature for a given species, is called the marker point. A convenient starting procedure is to perform the erosion of the body image (loss conversion), in order to expose the main limb axes, at the same time indicating their joint connections (bonds). The next step is to flag the selected markers at the coordinate system obtained and to draw the vectors showing the acceleration, velocity, and direction of motion. The procedure requires a considerable computing power of the computer system.

13.2 The problem of a parametrically incomplete observation

From the diagnostician's viewpoint, a quadroscope, dynamic image analysis is a very attractive option, since the acquisition of motion geometry parameters is based exclusively on the measurements of patient's relatively casual behavior, recorded by few cameras. The only nuisance are reflective (or glowing with their own light) markers fixed to the body. The drawback of this solution is a complete lack of information on the causative processes that lead to the motion of the limbs, as well as about the motion results, or, in other words, the kinematic impact of its action on the ground.

The elements trying to fill this gap were two types of contact posturometry systems, the first of which—strain gauges—collected information about the distribution of foot pressure on the ground (in the time domain) and the second, accelerometer analyzers, through a network of sensors located on the limbs, recorded multichannel motion speed and limb acceleration.

Since 2003, the Laboratory of Biotechnology ("LABIOT") began to construct measuring systems combining both of the above diagnostic features with the simultaneous measurement of the blood flow and even the electrical parameters of the skin and muscles (**Figure 1b**). Parameters registered by the prototypes, which were built according to the above schemes, initially showed little satisfactory repeatability; hence, a strict standardization (markering) of measurement points was introduced. The multithreaded systems corrected this way showed much greater stability and reproducibility of the results. Undeniably it was a technical success, yielding a large number of reliable measurement results interconnected on a common time base. As it turned out, technical success was only a partial solution to the problem

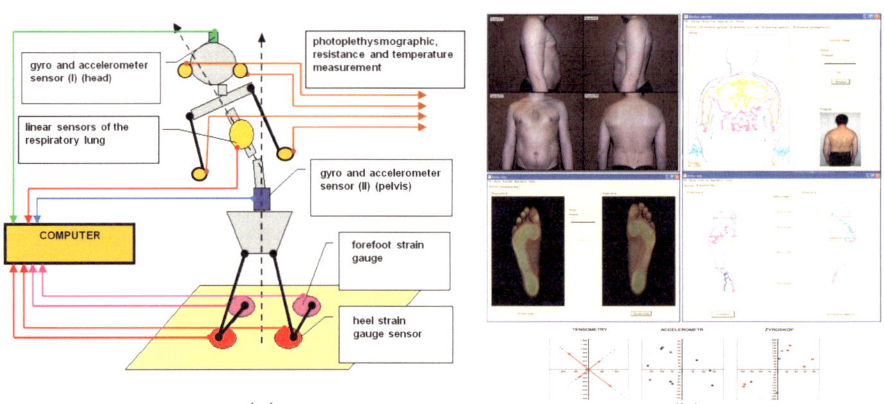

(a) (b)

Figure 1.
A scheme of the author's contact posturometry system: (a) built in 2003, which integrates a multichannel system of acceleration sensors with feet strain gauge sensors and sensors registering flows in the blood vessels; (b) numerically recording system (PSB1) quadroscopic-podoscopic, gyroscopic, accelerometric, and strain gauge parameters of the patient's body, presented at the Rehabilitation Fair in 2019.

and was not reflected in the clinical aspect, as numerous and accurately registered measurement parameters were presented in completely different physical units and thus were difficult to correlate and even more so to interpret [97, 98].

There is yet another question that arises—how to understand the body's dynamic state at various movement stages described by multiple numbers, whose calibration systems are anchored in completely different dimensions and physical units? Is there a common denominator that indicates the momentary, initial, and final status for such data distribution, let alone a reliable clinical interpretation?

Author details

Andrzej Jan Dyszkiewicz[1,2,3]* and Diana Hruby[1,2,3]

1 LABIOT Laboratory of Biotechnology, Cieszyn, Poland

2 Faculty of Mechatronics, University of Technology, Katowice, Poland

3 Academy of Physical Education, Katowice, Poland

*Address all correspondence to: andrzej@labio.pl

References

[1] Dietrich M, Kędzior K, Zagrajek T. Nieliniowa analiza deformacji układu kręgosłupa z uwzględnieniem stateczności. In: Nałęcz M, editor. Biocybernetyka i Inżynieria Biomedyczna 2004, Tom 5 – Biomechanika i Inżynieria Rehabilitacyjna. Warszawa: Akademicka Oficyna Wyd. "Exit"; 2004. pp. 259-276

[2] Arnold DH, Thompson M, Johnston A. Motion and position coding. Vision Research. 2007;**47**:2403-2410

[3] Blajer W, Dziewiecki K, Mazur Z. Aspekty modelowania dynamiki ruchu człowieka dla zadań symulacji odwrotnej. Acta Mechanica et Automatica. 2010;**4**(2):17-24

[4] Henderson L, Goodman ND, Tenenbaum JB, Woodward JF. The structure and dynamics of scientific theories, a hierarchical Bayesian perspective. Philosophy in Science. 2010;**77**(2):172-200

[5] Hossner EJ, Künzell S. Motorisches lernen. In: Mechling H, Munzert J, editors. Handbuch Bewegungswissenschaft-Bewegungslehre. Schorndorf: Hofmann; 2003

[6] Cisek P. Cortical mechanisms of action selection: The affordance competition hypothesis. Philosophical Transactions of the Royal Society B. 2007;**362**:1585-1599

[7] Kemp C, Tenenbaum JB, Niyogi SG. A probabilistic model of theory formation. Cognition. 2010;**114**(2):165-196

[8] Kettner RE et al. Prediction of complex two-dimensional trajectories by a cerebellar model of smooth pursuit eye movement. Journal of Neurophysiology. 1997;77:2115-2130

[9] Osiński W. Antropomotoryka. AWF: Poznań; 2003

[10] Pertyński W. Narzędzia topologiczne w opisie sterowania ruchami przez człowieka. Antropomotorika. 2006;**16**(35):77-87

[11] Szopa J, Mleczko E, Żak S. Podstawy Antropomotoryki. Warszawa – Kraków: PWN; 2000

[12] Pertyński W. Współczesne Teorie Uczenia się Ruchów i Sterowania Nimi u Człowieka. Katowice: Wyd. GWSH; 2008

[13] Polat U, Sagi D. The architecture of perceptual spatial interactions. Vision Research. 1994;**34**:73-78

[14] Poulet JFA, Hedwig B. Auditory orientation in crickets: Pattern recognition controls reactive steering. Proceedings of the National Academy of Sciences of the United States of America. 2005;**102**(15):665-669

[15] Seymour B. The maladaptive brain: Excitable pathways to chronic pain. Brain. 2012;**135**(2):316-318

[16] Seymour B, Daw ND, Roiser JP, Dayan P, Dolan R. Serotonin selectively modulates reward value in human decision-making. The Journal of Neuroscience. 2012;**32**(17):5833-5842

[17] Yoshida W, Seymour B, Koltzenburg M, Dolan RJ. Uncertainty increases pain: Evidence for a novel mechanism of pain modulation involving the periaqueductal gray. The Journal of Neuroscience. 2013;**33**(13):5638-5646

[18] Halanych KM. The new view of animal phylogeny. Annual Review of Ecology and Systematics. 2004;**35**:229-256

[19] Martindale MQ, Henry JQ. The development of radial and Biradial symmetry: The evolution

of bilaterality. American Zoologist. 1998;**38**(4):672-684

[20] Morawski JM. Bionics, modeling and computer simulation in investigation of basic human psycho-motor activities. Antropomotorika. 2009;**45**:91-108

[21] Tejszerska D, Świtoński E, Gzik M. Biomechanika Narządu Ruchu Człowieka. Radom: Wyd Nauk ITE-PIB; 2011

[22] Casey BJ et al. Dissociation of response conflict, attentional selection, and expectancy with functional magnetic resonance imaging. Proceedings of the National Academy of Sciences of the United States of America. 2000;**97**:8728-8733

[23] Daw ND, O'Doherty JP, Dayan P, Seymour B, Dolan RJ. Cortical substrates for exploratory decisions in humans. Nature. 2006;**441**(7095):876-879

[24] FitzGerald TH, Seymour B, Dolan RJ. The role of human orbitofrontal cortex in value comparison for incommensurable objects. The Journal of Neuroscience. 2009;**29**(26):8388-8395

[25] Giurfa M, Capaldi EA. Vectors, routes and maps: New discoveries about navigation in insects. Trends in Neurosciences. 1999;**22**:237-242

[26] Rushworth MF, Mars RB, Summerfield C. General mechanisms for making decisions? Current Opinion in Neurobiology. 2009;**19**:75-83

[27] Zatsiorsky VM. Kinetics of human motion. Theory and Practice of Physical Culture. 2002;**27**(8):24-28

[28] Yamaguchi GT. Dynamic modeling of musculoskeletal motion. A Vectorized Approach for Biomechanical Analysis in Three Dimensions. Boston: Kluwer Academic Publisher; 2001

[29] Miall RC, Wolpert DM. Forward models for physiological motor control. Neural Networks. 1996;**9**:1265-1279

[30] Mumford D. On the computational architecture of the neocortex. I. The role of the thalamo-cortical loop. Biological Cybernetics. 1991;**65**:135-145

[31] Nijhawan R, Wu S. Compensating time delays with neural predictions: Are predictions sensory or motor? Philosophical Transactions. Series A, Mathematical, Physical, and Engineering Sciences. 2009;**367**:1063-1078

[32] O'Doherty JP, Hampton A, Kim H. Model-based fMRI and its application to reward learning and decision making. Annals of the New York Academy of Sciences. 2007;**1104**:35-53

[33] Orban de Xivry J, Missal M, Lefèvre P. A dynamic representation of target motion drives predictive smooth pursuit during target. Journal of Vision. 2008;**8**:1-13

[34] Payne M, Hedwig B, Webb B. Multimodal predictive control in crickets. Lecture Notes in Computer Science. 2010;**6226**:167-177

[35] Seymour B, McClure SM. Anchors, scales and the relative coding of value in the brain. Current Opinion in Neurobiology. 2008;**18**(2):173-178

[36] Seymour B, O'Doherty JP, Dayan P, Koltzenburg M, Jones AK, Dolan RJ, et al. Temporal difference models describe higher-order learning in humans. Nature. 2004;**429**(6992):664-667

[37] Schmidt RA, Wrisberg CA. Motor Learning and Performance. Illinois, Champaign: Human Kinetics; 2004

[38] Summerfield C, Koechlin E. A neural representation of prior information during perceptual inference. Neuron. 2008;**59**:336-347

[39] Błaszczyk JW. Biomechanika Kliniczna. Katowice: AWF; 2010

[40] Konczak J. Neurophysiologische grundlagen der motorik. In: Mechling H, Munzert J, editors. Handbuch Bewegungswissenschaft-Bewegungslehre. Schorndorf: Hofmann; 2003

[41] Thura D, Cisek P. Deliberation and commitment in the premotor and primary motor cortex during dynamic decision-making. Neuron. 2014;**81**(6):1401-1416

[42] Thura D, Beauregard-Racine J, Fradet CW, Cisek P. Decision-making by urgency-gating: Theory and experimental support. Journal of Neurophysiology. 2012;**108**(11):2912-2930

[43] Haferlach T, Wessnitzer J, Mangan M, Webb B. Evolving a neural model of insect path integration. Adaptive Behavior. 2007;**15**:273-287

[44] Krashes M, DasGupta S, Vreede A, White B, Armstrong D, Waddell S. A neural circuit mechanism integrating motivational state with memory expression in Drosophila. Cell. 2009;**139**:416-427

[45] Menzel R, Giurfa M. Dimensions of cognition in an insect, the honeybee. Behavioral and Cognitive Neuroscience Reviews. 2006;**5**:24-40

[46] Menzel R et al. Honey bees navigate according to a map-like spatial memory. Proceedings of the National Academy of Sciences of the United States of America. 2005;**102**:3040-3045

[47] Heinze S, Homberg U. Map-like representation of celestial e-vector orientations in the brain of an insect. Science. 2007;**315**:995-997

[48] Schwarz S, Cheng K. Visual associative learning in two desert ant species. Behavioral Ecology and Sociobiology. 2010;**64**:2033-2041

[49] Webb B, Scutt T. A simple latency dependent spiking neuron model of cricket phonotaxis. Biological Cybernetics. 2000;**82**:247-269

[50] Collett M, Collett TS. How do insects use path integration for their navigation? Biological Cybernetics. 2000;**83**:245-259

[51] Lund HH, Webb B, Hallam J. Physical and temporal scaling considerations in a robot model of cricket calling song preference. Artificial Life. 1998;**4**:95-107

[52] Webb B. What does robotics offer animal behaviour? Animal Behaviour. 2000;**60**:545-558

[53] Kommu SS. Rehabilitation Robotics. Vienna: I-Tech Education and Publishing; 2007

[54] Vishal J. Design and control of active knee rehabilitation orthotic device (AKROD) [Mechanical Engineering Master's theses]. USA: Northeastern University of US; 2009

[55] Webb B. Transformation, encoding and representation. Current Biology. 2006;**16**:184-185

[56] Webb B, Reeve R. Reafferent or redundant: How should a robot cricket use an optomotor reflex? Adaptive Behavior. 2003;**11**:137-158

[57] Webb B. Neural mechanisms for prediction: Do insects have forward models? Trends in Neurosciences. 2004;**27**:278-282

[58] Wehner R. Desert ant navigation: How miniature brains solve complex tasks. Journal of Comparative Physiology. A, Neuroethology, Sensory, Neural, and Behavioral Physiology. 2003;**189**(8):579-588

[59] Dickinson A. Associative learning and animal cognition. Philosophical Transactions of the Royal Society B: Biological Sciences. 2012;**367**(1603):2733-2742

[60] Schatz B, Chameron S, Beugnon G, Collett TS. The use of path integration to guide route learning in ants. Nature. 1999;**399**:769-772

[61] Dyszkiewicz A. Standaryzacja diagnostyki narządu ruchu. In: Klukowski Krzysztof Medycyna Sportowa. Warsaw, Poland: Med Trib; 2012. pp. 51-55

[62] Dyszkiewicz A. Perspektywy wykorzystania skojarzonego systemu symetrii pomiarowych SSSP w rehabilitacji. In: Klukowski Krzysztof Medycyna Sportowa. Warsaw, Poland: Med Trib; 2012. pp. 56-61

[63] Dyszkiewicz A. Wpływ Elektrofonoforezy Levamisolu na Wyznaczniki Symetrii Ciała i Stężenie Kompleksów Immunologicznych w Wybranych Zespołach Bólowych Okolicy Lędźwiowo-Krzyżowej Kręgosłupa. Monografia Nr. 2353. Katowice: Uniwersytet Śląski; 2005. pp. 1-182

[64] Zielosko B, Dyszkiewicz A. Intelligent medical systems on internet technologies platform. In: Advances in Soft Computing, Monitoring, Security and Rescue Techniques in Multiagent Systems. Berlin/Heidelberg/New York: Springer; 2005. pp. 589-594. ISBN: 3-540-23245-1

[65] Dyszkiewicz A. Chapter 2: Local use of electrophonophoresis in the treatment of patients with immunologically based neuro-vertebral conflicts. In: Montenegro ML, editor. Pelvic Pain: Causes, Symptoms and Treatments. USA: Nova Science Publishers Inc.; 2012. pp. 45-76

[66] Cisek P. Integrated neural processes for defining potential actions and deciding between them: A computational model. The Journal of Neuroscience. 2006;**26**(38):9761-9770

[67] Cisek P, Kalaska JF. Neural correlates of reaching decisions in dorsal premotor cortex: Specification of multiple direction choices and final selection of action. Neuron. 2005;**45**(5):801-814

[68] Petryński W, Bernsztejn NA. O zręczności i jej rozwoju; uczenie się ruchów. Antropomotorika. 2005;**30**:71-86

[69] Daw ND, Gershman SJ, Seymour B, Dayan P, Dolan RJ. Model-based influences on humans' choices and striatal prediction errors. Neuron. 2011;**69**(6):1204-1215

[70] Bar M. The proactive brain: Using analogies and associations to generate predictions. Trends in Cognitive Sciences. 2007;**11**:280-289

[71] Cos I, Medleg F, Cisek P. The modulatory influence of endpoint controllability on decisions between actions. Journal of Neurophysiology. 2012;**108**(6):1764-1780

[72] Collett M, Collett T, Wehner R. Calibration of vector navigation in desert ants. Current Biology. 1999;**9**:1031-1034

[73] Cisek P, Puskas GA, El-Murr S. Decisions in changing conditions: The urgency-gating model. The Journal of Neuroscience. 2009;**29**(37):11560-11571

[74] Chittka L, Rossiter S, Skorupski P, Fernando C. What is comparable in comparative cognition. Philosophical Transactions of the Royal Society of London. Series B, Biological Sciences. 2012;**367**(1603):2677-2685

[75] Komendziński T. Multimoodalna dynamika koordynacji. AVANT. 2012;**III**(2):334-343

[76] Cos I, Bélanger N, Cisek P. The influence of predicted arm biomechanics on decision-making. Journal of Neurophysiology. 2011;**105**(6):3022-3033

[77] De Martino B, Kumaran D, Seymour B, Dolan RJ. Frames, biases, and rational decision-making in the human brain. Science. 2006;**313**(5787):684-687

[78] Grush R. The emulation theory of representation: Motor control, imagery, and perception. The Behavioral and Brain Sciences. 2003;**27**:377-442

[79] Kalisch R, Korenfeld E, Stephan KE, Weiskopf N, Seymour B, Dolan RJ. Context-dependent human extinction memory is mediated by a ventromedial prefrontal and hippocampal network. The Journal of Neuroscience. 2006;**26**(37):95

[80] Kording KP, Tenenbaum JB, Shadmehr R. The dynamics of memory are a consequence of optimal adaptation to a changing body. Nature Neuroscience. 2007;**10**(6):779-786

[81] Latash ML. Equilibrium-point hypothesis and internal inverse models. In: Raczek J, Waśkiewicz Z, Juras G, editors. Current Research in Motor Control. Katowice: AWF; 2000. pp. 44-49

[82] Winter DA. Biomechanics and Motor Control of Human Movement. USA: John Wiley & Sons; 2005

[83] Zago M, McIntyre J, Senot P, Lacquaniti F. Visuo-motor coordination and internal models for object interception. Experimental Brain Research. 2009;**192**(4):571-604

[84] Zakynthynaki MS, Stirling JR. Modelowanie matematyczne w biomechanice sportu. Sport Wyczynowy. 2003;**11-12**:467-468

[85] Cameron EL, Tai JC, Carrasco M. Covert attention affects the psychometric function of contrast sensitivity. Vision Research. 2002;**42**:949-967

[86] Bolles RC, Fanselow MS. A perceptual-defensive-recuperative model of fear and pain. The Behavioral and Brain Sciences. 1980;**3**(2):291-301

[87] Syczewska M, Öberg T. Spinal segmental movement changes during treadmill gait after stroke. Journal of Human Kinetics. 2006;**16**:39-56

[88] Lindstedt Ł, Rodzewicz C, Rzymkowski K, Kędzior. Experimental study of impact phenomena in case of a composite glider. In: XXIX OSTIV Congress; Berlin: Lüsse; 2008

[89] Lindstedt Ł. Badania eksperymentalne dotyczące bezpieczeństwa biernego pilota szybowca PW-5 podczas zderzenia z przeszkodą naziemną. Naukainnov Technologies. 2007;**1**:12-13

[90] Dyszkiewicz A, Opara J, Wróbel Z. Computerised system for evaluation of asymmetry of postural parameters coefficients in scoliosis. In: Fred A, Filipe J, Gamboa H, editors. Biomedical Engineering Systems and Technologies Communications in Computer and Information Science. Vol. 25, XIV. Berlin, Germany: Springer; 2009. pp. 277-280

[91] Dyszkiewicz A, Połeć P, Zajdel J, Chachulski D. Specific evaluation of pelvic radiograms and hip BMD in structural scoliosis reflectorica and reactive pain conditions of the backbone. In: Advances in Soft Computing, Information Technologies in Biomedicine. Berlin, Germany: Springer; 2010. pp. 491-507

[92] Dyszkiewicz A, Opara J. Znaczenie posturometrii i stabilografii w rehabilitacji i zapobieganiu upadkom u osób po udarze mózgu. In: Zeszyty

Metodyczno-Naukowe: Nowoczesna Rehabilitacja po Udarze Mózgu. Katowice: AWF; 2006. pp. 127-140

[93] Dyszkiewicz A. Wpływ Elektrofonoforezy Levamisolu na Wyznaczniki Symetrii Ciała i Stężenie Kompleksów Immunologicznych w Wybranych Zespołach Bólowych Okolicy Lędźwiowo-Krzyżowej Kręgosłupa. Monograficzna Praca Naukowa Nr. 2353. Katowice: Wydana Przez Uniwersytet Śląski; 2005

[94] Dyszkiewicz A, Gaździk T. Zastosowanie leków immunomodulacyjnych w leczeniu lędźwiowych konfliktów krążkowo-korzeniowych o dwutorowej symptomatologii. Ortopedia, Traumatologia, Rehabilitacja. 2004;**6**(2):190-197

[95] Dyszkiewicz A, Gaździk T, Opara J. Monitorowanie leczenia zespołów bólowych kręgosłupa lędźwiowego z zastosowaniem niesterydowego leku przeciwzapalnego i żelu na bazie naturalnego komponentu olejkowo-eterycznego. Ortopedia, Traumatologia, Rehabilitacja. 2006;**2**(6):210-218

[96] Jamtvedt G, Young JM, Kristoffersen DT, Thomson O'Brien MA, Oxman AD. Audit and Feedback: Effects on Professional Practice and Health Care Outcomes. USA: John Wiley & Sons; 2003. pp. 1-259

[97] Herbert R, Jam Tvedt G, Mead J, Burger HK. Practical Evidence-Based Physiotherapy. Amsterdam, Netherlands: Elsevier; 2003. pp. 1-186

[98] Bury T, Mead J. Evidence-Based Healthcare: A Practical Guide for Therapists. Oxford: Butterworth-Heinemann; 1995. pp. 1-264

Section 3

Applied Biomechanics

Chapter 5

Biomechanical Model Improving Alzheimer's Disease

Eliete Biasotto Hauser, Wyllians Vendramini Borelli and Jaderson Costa da Costa

Abstract

The aim this study is to describe the algorithms of kinetic modeling to analyze the pattern of deposition of amyloid plaques and glucose metabolism in Alzheimer's dementia. A two-tissue reversible compartment model for Pittsburgh Compound-B ($[^{11}C]PIB$) and a two-tissue irreversible compartment model for $[^{18}F]$2-fluoro-2-deoxy-D-glucose ($[^{18}F]FDG$) are solved applying the Laplace transform method in a system of two first-order differential equations. After calculating a convolution integral, the analytical solutions are completely described. In order to determine the parameters of the model, information on the tracer delivery is needed. A noninvasive reverse engineer technique is described to determine the input function from a reference region (carotids and cerebellum) in PET image processing, without arterial blood samples.

Keywords: noninvasive input function, Laplace transform, kinetic modeling, radiotracer, positron emission tomography (PET), reference region, region of interest, time activity curve

1. Introduction

Positron emission tomography (PET) [1, 2], is a functional imaging technology that visualizes physiological changes through the administration of radiopharmaceutical molecular tracers into living systems. PET with measures the local concentration of a tracer in the region of interest (ROI) or target tissue.

PET with $[^{11}C]PIB$ and $[^{18}F]FDG$ radiotracers are widely used in the clinical setting for patients with neurodegenerative diseases like Alzheimer's disease. The $[^{18}F]FDG$-PET indirectly measures neuronal metabolism, subsequently allowing the identification of brain regions with increased or decreased activity. Individuals with progressive amnestic dementia show a specific pattern of FDG uptake that distinguishes their brains from other types of pathologies. Thus, this technique directly impacts the treatment selected for this patient. However, this technique is still under study to improve its accuracy power and to decrease patient discomfort undergoing this diagnostic tool.

Cognitive aging is also a subject of interest of PET studies. This technique can be used to investigate abnormal binding occurs in clinically normal individuals, prior to the development of cognitive changes. Higher binding in nondemented subjects suggests that $[^{11}C]PIB$ amyloid imaging may be sensitive for detection of a preclinical Alzheimer's disease state. Age-related cognitive changes impact the brain functioning and subsequently neuronal activity. Frontal and medial temporal regions are particularly vulnerable for the aging process. Nonetheless, a group of elderly named SuperAgers exhibit exceptional memory ability and a specific brain signature [3–7]. SuperAgers appear to maintain neuronal activity throughout the aging process, showing stable neuronal activity in the frontal lobe when compared with normal agers.

Mathematical modeling seeks to describe the processes of distribution and elimination through compartments, which represent different regions (for example, the vascular space, interstitial, intracellular) or different chemical stages. Noninvasive methods have been used successfully in PET image studies [8–13].

In order to determine the parameters of the model, information on the tracer delivery is needed in the form of the input function that represents the time-course of tracer concentration in the arterial blood or plasma is non-invasively obtained by non-linear regression [14], from the time-activity curve in a reference region (carotids and cerebellum).

The Laplace transform is used to generate the exact solution solutions of the $[^{11}C]PIB$ two-tissue reversible compartment model, [15], and $[^{18}F]FDG$ two-tissue irreversible compartment model, [16, 17], applying the Laplace transform method in a system of two first order differential equations. From a reference region (carotids and cerebellum) the technique allows to estimate the concentration in each compartment of the region of interest, illustrated in **Figure 1**.

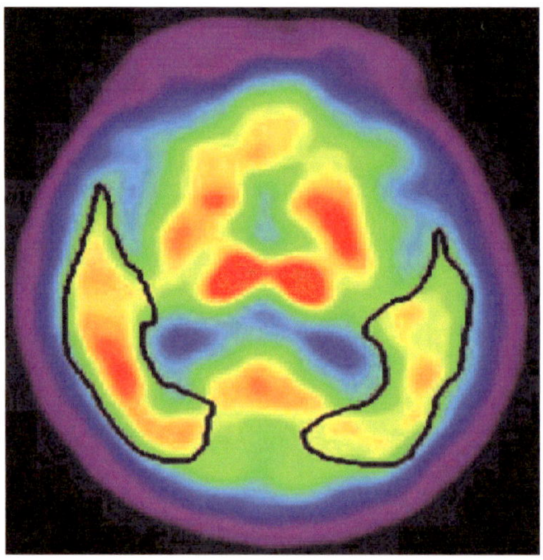

Figure 1.
Region of interest outlined in temporal areas bilaterally in an axial slice.

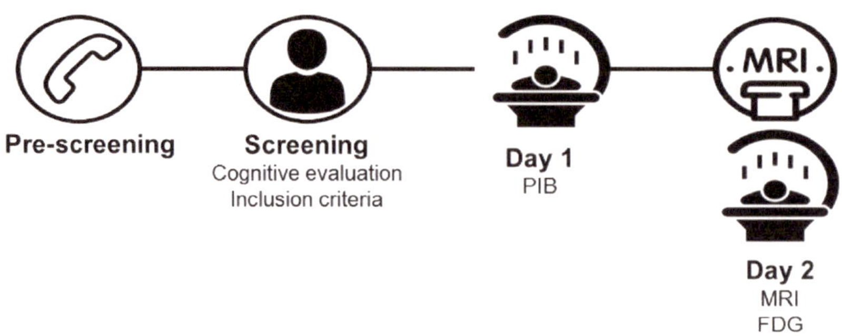

Figure 2.
SuperAgers project study protocol.

1.1 Image analysis and data generation

Data used in this work was obtained with $[^{18}F]FDG$ and $[^{11}C]PIB$ synthesized by the Cyclotron at the Instituto do Cérebro (InsCer/BraIns) at the Pontifical Catholic University of Rio Grande do Sul PUCRS, in studies (**Figure 2**) approved by medical ethics committee with *PET/CT* imaging.

Using software PMOD, the 3D Gaussian pre-processing tool is used to make decay, attenuation, scatter and dead time corrections.

1.2 Effective dose injected (EDI) and half-life

According [1], the effective dose injected can be calculated as:

$$EDI = C_a^i e^{-\frac{\ln 2}{t_{1/2}}(t_0 - t_i)} - C_a^e e^{-\frac{\ln 2}{t_{1/2}}(t_e - t_0)} \tag{1}$$

where C_a^i is the dose measured before injection at time t_i, C_a^i is the rest dose after injection measured at time t_e and $t_{1/2}$ is the half-life of the tracer.

The radioactivity of $[^{11}C]PIB$ decays with a half-life of $t_{1/2}$ = 20 min and of $[^{18}F]FDG$ with a half-life of 109.7 min.

2. Models compartments

Mathematical modeling seeks to describe the processes of distribution and elimination through compartments, which represent different regions (for example, the vascular space, interstitial, intracellular) o different chemical stages.

Transferring rate from one compartment to another, is proportional to concentration in the compartment of origin. Compartmental model is an important kinetic modeling technique used for quantification of PET. Each compartment is characterized by the concentration within it as a function of time. The physiological and metabolic transport processes are described mathematically by the analysis of mass balance equations.

A compartment model is represented by a system of differential equations, where each equation represents the sum of all the transfer rates to and from a specific compartment:

$$\frac{d}{dt} C_i(t) = \sum_{j=1, j\neq i}^{N} \left[K_{ij} C_j(t) - K_{ji} C_i(t)\right], \tag{2}$$

where $C_i(t)$ is the concentration of radioactive tracer in compartment i, N is the number of sections of the model, K_{ji} is the rate constant for transfer from compartment j to compartment i.

Figure 3 illustrates a reversible compartment model, that is be used to investigated the $[^{11}C]PIB$ metabolism, [15], because this tracer enters a reference region and since it does not find amyloid plaques to bind it leaves rapidly.

The irreversible two compartment model (**Figure 2** with $k_4 = 0$) is used for description of tracer $[^{18}F]FDG$, [16, 17], which first enter a free compartment, C1, and is then metabolized irreversibly in the second compartment C2.

In order to determine the parameters of the model, it is necessary to have information about the tracer delivery in the form of an input function representing the time-course of tracer concentration in arterial blood or plasma.

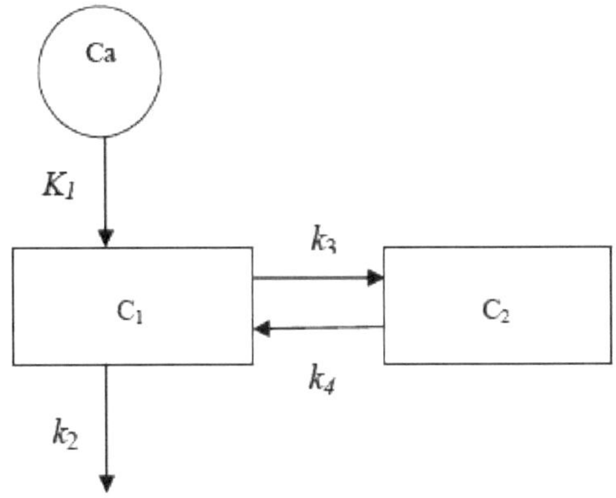

Figure 3.
A schematic diagram of a reversible two compartments model to illustrate the flux of tracer between blood (Ca) and and two tissues (C1 and C2).

2.1 Estimation of rate constants

In order to estimate the parameters k_{ij}, k_{ji}, a nonlinear regression problem is solved using the Levenberg-Marquardt method [18, 19]. The sensitivity equations are generated partially deriving Eq. (2) with respect to the parameters k_{ij}, k_{ji}

$$\begin{cases} \dfrac{\partial}{\partial K_{ij}}\left(\dfrac{d}{dt}C_i(t) = \sum\limits_{j=1, j\neq i}^{N}\left[K_{ij}C_j(t) - K_{ji}C_i(t)\right]\right) \\ \dfrac{\partial}{\partial K_{ji}}\left(\dfrac{d}{dt}C_i(t) = \sum\limits_{j=1, j\neq i}^{N}\left[K_{ij}C_j(t) - K_{ji}C_i(t)\right]\right) \end{cases} \tag{3}$$

Over which region of interest (ROI) is defined discrete TAC using the image processing. The Jacobian matrix it consists of the column vectors whose values resulting from the numerical integration of the sensitivity equations with respect to time.

3. Two-tissue reversible compartment model of $[^{11}C]PIB$

The mathematical model for the representation of the dynamics of $[^{11}C]PIB$ radiotracer, [11, 15, 20], is expressed by the system of two differential equations:

$$\begin{cases} \dfrac{dC_1}{dt} = K_1 C_a(t) - (k_2 + k_3)C_1(t) + k_4 C_2(t) \\ \dfrac{dC_2}{dt} = k_3 C_1(t) - k_4 C_2(t) \\ C_1(0) = 0, C_2(0) = 0. \end{cases} \tag{4}$$

where $C_a(t)$ is the arterial input function (AIF) considered to be known, $C_1(t)$ and $C_2(t)$ are, respectively, the concentration within the non-displaceable and displaceable compartments and K_1 and k_2, k_3, k_4 are kinetic rate constants which have to be determined.

The Laplace transform with respect to t is applied to solve the system of differential equations Eq. (4), with the notation

$$£\{C_i(t)\} = \overline{C}_i(s) = \int_0^{\infty} e^{-st} C_i(t) dt$$

and

$$£\left\{\frac{dC_k(t)}{dt}\right\} = s\overline{C}_i(s) - C_i(0).$$

An algebraic system is obtained

$$\begin{cases} (s + k_2 + k_3)\overline{C}_1(s) - k_4\overline{C}_2(s) = K_1\overline{C}_a(s) \\ -k_3\overline{C}_1(s) + (s + k_4)\overline{C}_2(s) = 0 \end{cases}, \quad (5)$$

that can be written in matrix form as

$$\begin{bmatrix} s + k_2 + k_3 & -k_4 \\ -k_3 & s + k_4 \end{bmatrix} \begin{bmatrix} \overline{C}_1(s) \\ \overline{C}_2(s) \end{bmatrix} = \begin{bmatrix} K_1\overline{C}_a(s) \\ 0 \end{bmatrix}. \quad (6)$$

The solution of the algebraic system (6) is

$$\begin{bmatrix} \overline{C}_1(s) \\ \overline{C}_2(s) \end{bmatrix} = \begin{bmatrix} s + k_2 + k_3 & -k_4 \\ -k_3 & s + k_4 \end{bmatrix}^{-1} \begin{bmatrix} K_1\overline{C}_a(s) \\ 0 \end{bmatrix}. \quad (7)$$

The inverse matrix is

$$\begin{bmatrix} s + k_2 + k_3 & -k_4 \\ -k_3 & s + k_4 \end{bmatrix}^{-1} = \frac{1}{(s + k_2 + k_3)(s + k_4) - k_3k_4} \begin{bmatrix} s + k_4 & k_4 \\ k_3 & s + k_2 + k_3 \end{bmatrix}. \quad (8)$$

Therefore,

$$\begin{aligned} \overline{C}_1(s) &= \frac{(s + k_4)K_1\overline{C}_a(s)}{s^2 + (k_2 + k_3 + k_4)s + k_2k_4} \\ \overline{C}_2(s) &= \frac{k_3K_1\overline{C}_a(s)}{s^2 + (k_2 + k_3 + k_4)s + k_2k_4} \end{aligned} \quad (9)$$

Using the inverse Laplace in Eq. (9), results

$$\begin{aligned} C_1(t) &= £^{-1}\left\{\frac{(s + k_4)K_1\overline{C}_a(s)}{s^2 + (k_2 + k_3 + k_4)s + k_2k_4}\right\} \\ C_2(t) &= £^{-1}\left\{\frac{k_3K_1\overline{C}_a(s)}{s^2 + (k_2 + k_3 + k_4)s + k_2k_4}\right\}. \end{aligned} \quad (10)$$

Now, the proprieties inverse Laplace transform are used, considering * to denote the convolution[1].

[1] The property of commutativity is valid in convolution operation for Laplace transform of $f(t)$ and $g(t)$ functions, defined by $f(t) * g(t) = \int_0^t f(u)g(t-u)du = \int_0^t f(t-u)g(u)du$.

$$C_1(t) = \pounds^{-1}\left\{\frac{(s+k_4)}{s^2+(k_2+k_3+k_4)s+k_2k_4}\right\} * K_1\pounds^{-1}\{\overline{C}_a(s)\}$$
$$C_2(t) = \pounds^{-1}\left\{\frac{k_3}{s^2+(k_2+k_3+k_4)s+k_2k_4}\right\} * K_1\pounds^{-1}\{\overline{C}_a(s)\}. \tag{11}$$

Then,

$$C_1(t) = K_1\pounds^{-1}\left\{\frac{s+k_4}{s^2+(k_2+k_3+k_4)s+k_2k_4}\right\} * C_a(t)$$
$$C_2(t) = K_1\pounds^{-1}\left\{\frac{k_3}{s^2+(k_2+k_3+k_4)s+k_2k_4}\right\} * C_a(t). \tag{12}$$

$$\frac{s+k_4}{s^2+(k_2+k_3+k_4)s+k_2k_4} = \frac{A}{s-s_1} + \frac{B}{s-s_2}$$
$$\frac{k_3}{s^2+(k_2+k_3+k_4)s+k_2k_4} = \frac{C}{s-s_1} + \frac{D}{s-s_2} \tag{13}$$

In Eq. (13) s_1 and s_2 are the roots of $s^2 + (k_2 + k_3 + k_4)s + k_2k_4 = 0$, dependent on transport constants k_2, k_3, and k_4. The parameters A, B, C, and D are obtained by partial fraction decomposition technique. Then, because that the inverse Laplace transforms are simply linear combinations of exponential functions with the exponents s_1 and s_2 depending on k_2, k_3, and k_4. Applying the linearity property of the inverse Laplace transform is obtained

$$\begin{aligned} C_1(t) &= K_1\{Ae^{s_1t} + Be^{s_2t}\} * C_a(t) \\ C_2(t) &= K_1\{Ce^{s_1t} + De^{s_2t}\} * C_a(t) \end{aligned}. \tag{14}$$

The analytical solution of the reversible two-compartment model for $[^{11}C]PIB$ (4) is

$$\begin{aligned} C_1(t) &= K_1\left\{Ae^{s_1t}\int_0^t e^{-s_1u}C_a(u)du + Be^{s_2t}\int_0^t e^{-s_2u}C_a(u)du\right\} \\ C_1(t) &= K_1\left\{Ce^{s_1t}\int_0^t e^{-s_1u}C_a(u)du + De^{s_2t}\int_0^t e^{-s_2u}C_a(u)du\right\} \end{aligned}. \tag{15}$$

In Eq. (15), it is visible the importance of construction of input function $C_a(t)$ in order to make it possible to calculate the integral

$$I = \int_0^t e^{s_i u} C_a(u) du. \tag{16}$$

4. Two-tissue irreversible compartment model of $[^{18}F]FDG$

$[^{18}F]FDG$ is a glucose analogue used to evaluate brain's metabolic activity in vivo through positron emission tomography with computed tomography (PET/CT). The irreversible two-compartment model for $[^{18}F]FDG$ is used for description of this tracer, which is first entering a free compartment, C1, and is then metabolized irreversibly in the second compartment C2, [16].

$$\begin{cases} \frac{d}{dt}C_1(t) = K_1C_a(t) - (k_2 + k_3)C_1(t) \\ \frac{d}{dt}C_2(t) = k_3C_1(t) \\ C_1(0) = 0, C_2(0) = 0 \end{cases}, \tag{17}$$

where $C_a(t)$ is the input function and is considered to be known, $C_1(t)$ and $C_2(t)$ are the concentration in C1 and C2 compartments, respectively, and K_1, k_2, k_3 are positives proportionality rates describing, the tracer influx into and the tracer outflow from the compartment (transport constants).

Similarly to that developed in the previous section, considering $k_4 = 0$, applying the Laplace transform with respect to t in Eq. (17), appear the algebraic system

$$\begin{cases} (s + k_2 + k_3)\overline{C}_1(s) = K_1\overline{C}_a(s) \\ -k_3\overline{C}_1(s) + s\overline{C}_2(s) = 0 \end{cases}. \tag{18}$$

Eq. (18) is represented matrically

$$\begin{bmatrix} \overline{C}_1(s) \\ \overline{C}_2(s) \end{bmatrix} = \begin{bmatrix} s + k_2 + k_3 & 0 \\ -k_3 & s \end{bmatrix}^{-1} \begin{bmatrix} K_1\overline{C}_a(s) \\ 0 \end{bmatrix}. \tag{19}$$

$$\begin{bmatrix} \overline{C}_1(s) \\ \overline{C}_2(s) \end{bmatrix} = \frac{1}{s(s + k_2 + k_3)} \begin{bmatrix} s & 0 \\ k_3 & s + k_2 + k_3 \end{bmatrix} \begin{bmatrix} K_1\overline{C}_a(s) \\ 0 \end{bmatrix}. \tag{20}$$

Then,

$$\begin{aligned} \overline{C}_1(s) &= \frac{K1\overline{C}_a(s)}{s + k_2 + k_3} \\ \overline{C}_2(s) &= \frac{k_3K_1\overline{C}_a(s)}{(s + k_2 + k_3)s} = \frac{k_3C_1(s)}{s} \end{aligned} \tag{21}$$

$$\begin{aligned} C_1(t) &= \pounds^{-1}\left\{\frac{K_1\overline{C}_a(s)}{(s + k_2 + k_3)}\right\} = K_1\pounds^{-1}\left\{\frac{1}{(s + k_2 + k_3)}\right\} * \pounds^{-1}\{\overline{C}_a(s)\} \\ C_2(t) &= \pounds^{-1}\left\{\frac{k_3\overline{C}_1(s)}{s}\right\} = k_3 * \pounds^{-1}\{\overline{C}_1(s)\}. \end{aligned} \tag{22}$$

The representation Eq. (22) implies that

$$\begin{aligned} C_1(t) &= K_1e^{-(k_2+k_3)t} * C_a(t) = K_1\int_0^t e^{-(k_2+k_3)(t-u)}C_a(u)du \\ C_2(t) &= k_3 * C_1(t) = k_3\int_0^t C_1(u)du. \end{aligned} \tag{23}$$

Then, with $k_2 + k_3 > 0$, the analytical solution of the irreversible two compartment model for $[^{18}F]FDG$ Eq. (17) is

$$\begin{aligned} C_1(t) &= K_1e^{-(k_2+k_3)t}\int_0^t e^{(k_2+k_3)u}C_a(u)du \\ C_2(t) &= k_3\int_0^t C_1(u)du. \end{aligned} \tag{24}$$

It is important now to choose a suitable model to represent the input function $C_a(t)$, which makes it possible to calculate the integral $\int_0^t e^{(k_2+k_3)u} K_1 C_a(u)du$.

4.1 The input concentration

The knowledge of the input function is mandatory in quantifying by compartmental kinetic modeling. The radioactivity concentration of arterial blood can be measured during the course of the scan collecting blood samples.

Several techniques have been proposed for obtaining input function. [9] present five different forms to measure this data and [8] eight methods for the estimation image input function in dynamic $[^{18}F]FDG$ PET human brain. The image arterial input function provides data that are similar to arterial blood input methods and can be used to quantify, noninvasively, in PET studies, according to previous studies [8, 10, 13, 15, 20]. This technique calculate the input function using linear and nonlinear regression applied in a applied to a discrete set of data, discrete time activity curve (TAC) of reference region [11].

4.2 Input function derived of PET image

The dynamics of the radiotracer, [11, 17], on the reference region is governed by the differential equation

$$\frac{dC_r}{dt} = K_1' C_a(t) - k_2' C_r(t) \tag{25}$$

where $C_a(t)$ is the concentration of the radiotracer in the arterial blood, $C_r(t)$ is the concentration of the radiotracer in the reference region and $K_1' > 0$ and $k_2' > 0$ are the proportionality rates describing, respectively, the tracer influx into and the outflow from the reference tissue.

$C_r(t)$ is constructed from a TAC of a reference region [11].

After this, deriving $C_r(t)$ we obtain $C_a(t)$, which is the AIF, using

$$C_a(t) = \frac{1}{K_1'}\frac{dC_r}{dt} + \frac{k_2'}{K_1'} C_r(t) \tag{26}$$

The transport of the radiotracer across of arterial blood is very fast in the first few minutes and then decreases slowly. Then, it may be appropriate to estimate the $C_r(t)$ in a few stages as piecewise function, [16]. This is defined for three stages in the equation

$$\begin{aligned} C_r(t) = {} & (H(t-t_0) - H(t-t_1))C_{rf}(t) + (H(t-t_1) - H(t-t_2))C_{rI}(t) \\ & + H(t-t_2)C_{rs}(t), \end{aligned} \tag{27}$$

where $C_{rf}(t)$, $C_{rI}(t)$ and $C_{rs}(t)$ are the concentration of the radiotracer on the reference region, respectively, for the fast, intermediate and slow stage. $H(t)$ is the Heaviside function defined by

$$H(t-a) = \begin{cases} 0, t < a, \\ 1, t \geq a. \end{cases} \tag{28}$$

$$H(t-a) - H(t-b) = \begin{cases} 0, t < a \; and \; t \geq b, \\ 1, a \leq t < b. \end{cases} \tag{29}$$

DOI: http://dx.doi.org/10.5772/intechopen.92047

5. Results and discussion

In order to obtain the analytical solution of two-compartment model, Eq. (15) for $[^{11}C]PIB$ and Eq. (24) for $[^{18}F]FDG$ radiotracer, the important step is to determine $C_r(t)$ that will allow you to calculate the input function (Eq. (26)). In the reference region, $C_r(t)$, is approximated by means of linear and nonlinear regression of the data obtained from a discrete TAC curve on a positron emission tomography (PET) image, using PMOD, a biomedical image quantification software.

5.1 $C_r(t)$ for $[^{11}C]PIB$ radiotracer

For $[^{11}C]PIB$ tracer is chosen as reference region the left and right cerebellum, known to be amyloid free illustrated in **Figure 4**. The left and right cerebellum are

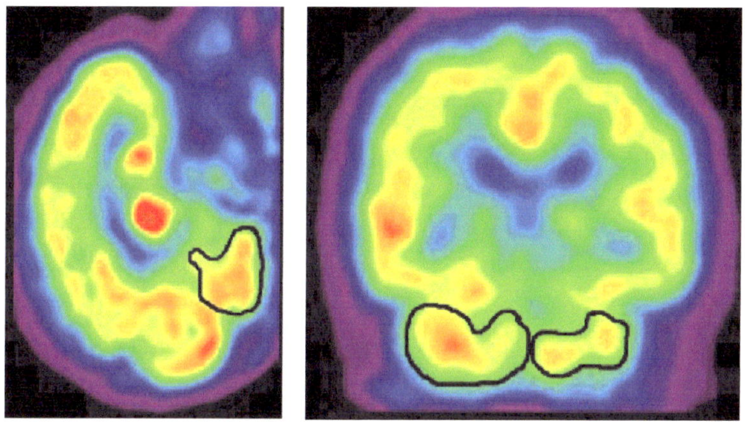

Figure 4.
Region of reference outlined in both cerebellar gray matter in a sagittal (left) and coronal (right) slices.

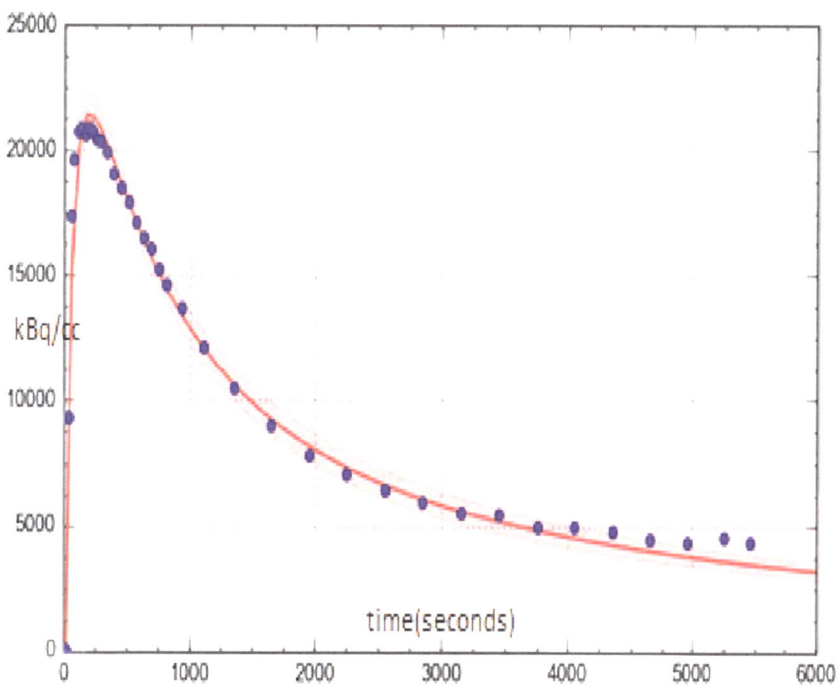

Figure 5.
Discrete and fitted rational cerebellum TAC ($C_r(t)$): the best model.

clearly visible, defined as the discrete left TAC and right TAC, and then the discrete average TAC is generated.

Nonlinear regression is applied to determine the parameters of the model chosen to approximate $C_r(t)$, from a discrete TAC curve.

The rational model (**Figure 5**) showed to be adequate for those seven patients (**Figure 6**), only differentiating by the values of the parameters a, b, c, d.

$$C_r(t) = \frac{at + b}{ct^2 + dt + 1} \tag{30}$$

The technique was applied for 7 patients considering the activity for the reference regions: right cerebellum, left cerebellum, and also for the mean of both (total of 24 simulations). The minor R^2 was 0.96421. The attempt to use an average of all discrete TAC was not to be adequate.

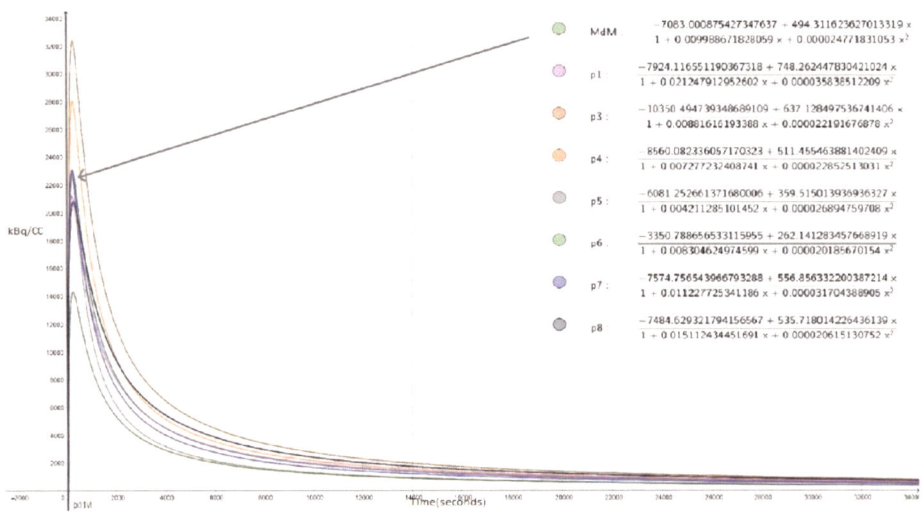

Figure 6.
Average (MdM) and rational fitted cerebellum time active (P1,P3-P8) curve ($C_r(t)$).

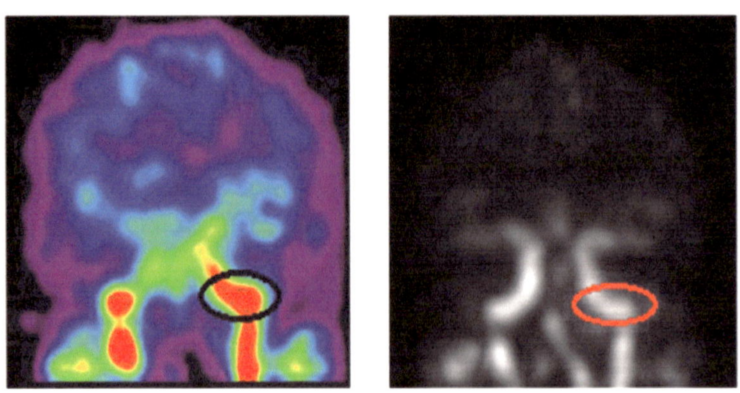

Figure 7.
Region of reference outlined in the left carotid artery region in a coronal slice in both PET (left) and MIP (right) images.

DOI: http://dx.doi.org/10.5772/intechopen.92047

5.2 $C_r(t)$ for $[^{18}F]FDG$ radiotracer

In order to obtain $C_r(t)$ for $[^{18}F]FDG$ radiotracer, the carotids are chosen as reference region. Manually, it is defined as volumes-of-interests (VOIs), illustrated in **Figure** 7, using a biomedical image quantification software PMOD. Over which the left and right carotid arteries where clearly visible, is defined discrete TAC.

Then, it may be appropriate to estimate the $C_r(t)$ considering the fast and slow stage.

After this, we apply regression techniques, and in two stages of the time, a good option that came up was the piecewise function logistical to describe the behavior of

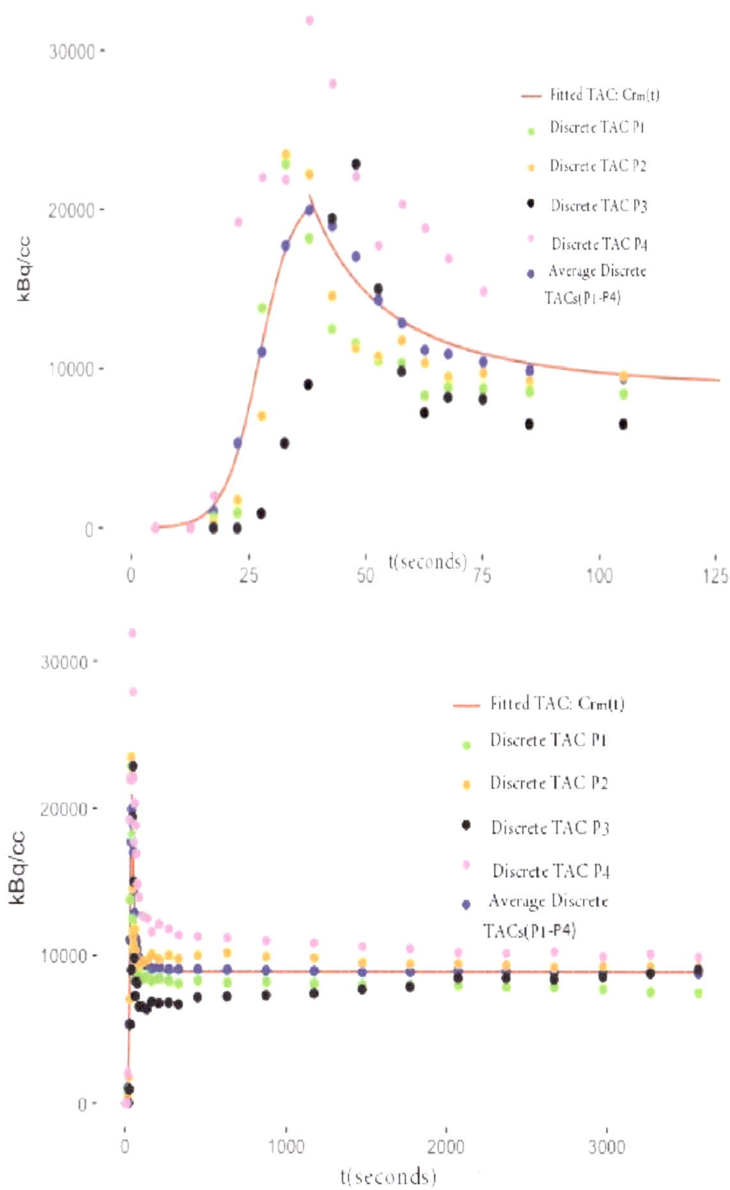

Figure 8.
Discrete TACs (P1–P4) and average logistic fitted carotid TAC ($C_{rm}(t)$) [21].

the mean of the discrete TACs of four patients (considering left volume), **Figure 8**, with correlation coefficient of 0.9947 (at least) is

$$C_r(t) = \frac{[H(t) - H(t - 37.5)]21088.3809}{1 + 1861.3158e^{0.2801t}} + \frac{[H(t - 37.5) - H(t - 3750)]8921.3}{1 - 1.7712e^{0.03t}} \tag{31}$$

It may be convenient in the diagnosis of Alzheimer's disease to consider the specific time interval seconds, [22, 23]. In this time interval, the graphs in **Figure 8** show the comparison between the values estimated by the function and the concentration of the FDG radiotracer in the left VOI. As we can see, the estimated values between 1170 and 2970 s were close to the original values, with the lowest average relative error is 0.0582 and the highest is 0.1096.

6. Conclusion

The aim this study was described the algorithms of kinetic modeling to analyze the pattern of deposition of amyloid plaques and glucose metabolism in Alzheimer's dementia, obtaining the exact solution of the $[^{11}C]PIB$ two-tissue reversible compartment model and a $[^{18}F]FDG$ two-tissue irreversible compartment model. Was solved a system of two first-order differential equations, applying the Laplace transform technique. Many sources of errors are involved in this problem. For example, the gathering data in image processing and the input function construction. With the exception of these errors, is assuring by using Laplace method proposed, there will not be error accumulation.

Longitudinal studies, without arterial blood samples, can assist in the calculation of the dose of medicine, providing the stabilization of cognitive impairment, behavior and the performance of activities of daily living. The technique here described can be used to analyze the pattern of deposition of amyloid plaques, glucose metabolism, the cortical and functional structure of the brain of SuperAgers in relation to cognitively normal elderly and individuals with Alzheimer's dementia. Older adults with exceptional memory ability are coined SuperAgers. Their preserved cognitive capacities with aging may help uncover neuromechanisms of dementia. These individuals showed whole-brain volume similar to middle-aged individuals and some areas thicker than usual agers. Intriguingly, they also exhibited decreased atrophy rate when compared to normal older adults. To our knowledge, their brain functional integrity is yet to be uncovered.

Thanks

This study was made possible by a team work from all members of the SuperAgers project: Lucas Porcello Schilling, Louise Mross Hartmann, Ana Maria Marques da Silva, Cristina Sebastiao Matushita, Mirna Wetters Portuguez, Alexandre Rosa Franco, and Ricardo Bernardi Soder. This research was partially supported by CNPq, project number 403029/2016-3 FAPERGS, project number 27971.414.15498.22062017.

Nomenclature

AIF	arterial input function
CT	computed tomography
EDI	effective dose injected
MIP	maximum intensity projection

PET	positron emission tomography
ROI	region of interest
TAC	time activity curve
VOI	volume of Interest

Author details

Eliete Biasotto Hauser[1,2]*, Wyllians Vendramini Borelli[2] and Jaderson Costa da Costa[2]

1 School of Technology, PUCRS, Porto Alegre, Brazil

2 Brain Institute of Rio Grande do Sul (BraIns), PUCRS, Porto Alegre, Brazil

*Address all correspondence to: eliete@pucrs.br

References

[1] Hsieh CH. Positron Emission Tomography Current Clinical and Research Aspects. Croatia: IntechOpen; 2012. pp. 336 DOI: 10.5772/1280

[2] Basu S, Zaidi H, Holm S, Abass A. Quantitative techniques in PET-CT imaging. Current Medical Imaging Reviews. 2011;**7**(3):216-233. DOI: 10.2174/157340511796411186

[3] Borelli WV, Schilling L, Radaelli G, Ferreira L, Pisani L, Portuguez M, et al. Neurobiological findings associated with high cognitive performance in older adults: A systematic review. International Psychogeriatrics. 2018; **30**(12):1813-1825. DOI: 10.1017/S1041610218000431

[4] Borelli WV, Carmona KC, Studart-Neto A, Nitrini R, Caramelli P, da Costa JC. Operationalized definition of older adults with high cognitive performance. Dementia & Neuropsychologia. 2018;**12**(3):221-227. DOI: 10.1590/1980-57642018dn12-030001

[5] Rogalski EJ, Gefen T, Shi J, Samimi M, Bigio E, Weintraub S, et al. Youthful memory capacity in old brains: Anatomic and genetic clues from the northwestern superaging project. Journal of Cognitive Neuroscience. 2013;**25**(1):29-36. DOI: 10.1162/jocn-a-00300

[6] Harrison T, Weintraub S, Mesulam M, Rogalski E. Superior memory and higher cortical volumes in unusually successful cognitive aging. Journal of the International Neuropsychological Society. 2012;**18**(6): 1081-1085. DOI: 10.1017/S1355617712000847

[7] Rodell AB, O'Keefe G, Rowe C, Villemagne V, Gjedde A. Cerebral blood flow and αβ-amyloid estimates by WARM analysis of [^{11}C]PiB uptake distinguish among and between neurodegenerative disorders and aging. Frontiers in Aging Neuroscience. 2017;**8**: 1-11. DOI: 10.3389/fnagi.2016.00321

[8] Zanotti-Fregonara P, Fadaili EM, Maroy R, Comtat C, Souloumiac A, Jan S, et al. Comparison of eight methods for the estimation of the image-derived input function in dynamic [^{18}F]-FDG PET human brain studies. Journal of Cerebral Blood Flow Metabolism. 2009;**29**(11):1825-1835. DOI: 10.1038/jcbfm.2009.9

[9] Laforest R, Sharp TL, Engelbach JA, Fetting NM, Herrero P, Kim J, et al. Measurement of input functions in rodents: Challenges and solutions. Nuclear Medicine and Biology. 2005;**3**: 679-685. DOI: 10.1016/j.nucmedbio.2005.06.012

[10] Zhou S, Chen K, Reiman EM, Li DM, Shan B. A method for generating image-derived input function in quantitative ^{18}F-FDG PET study based on the monotonicity of the input and output function curve. Nuclear Medicine Communications. 2012;**33**(4): 362-370. DOI: 10.1097/MNM.0b013e32834f262e

[11] Su Y, Blazey TM, Snyder AZ, Raichle ME, Hornbeck RC, Aldea P, et al. Quantitative amiloid imaging using image-derived arterial input function. PLoS One. 2015;**10**(4): e0122920. DOI: 10.1371/journal.pone.0122920

[12] Vriens D, de Geus-Oei LF, Oyen WJ, Visser EP. A curve-fitting approach to estimate the arterial plasma input function for the assessment of glucose metabolic rate and response to treatment. Journal of Nuclear Medicine. 2009;**50**(12):1933-1939. DOI: 10.2967/jnumed.109.065243

[13] Chen K, Bandy D, Reiman E, Huang SC, Lawson M, Feng D, et al. Noninvasive quantification of the cerebral metabolic rate for glucose using positron emission tomography, ^{18}F-Fluoro-2-Deoxyglucose, the Patlak method, and an image-derived input function. Journal of Cerebral Blood Flow & Metabolism. 1998;**18**(7):716-723. DOI: 10.1097/00004647-199807000-00002K

[14] Bates DM, Watts DG. Nonlinear Regression and its Applications. New York: Wiley; 1988. pp. 365. DOI: 10.1002/9780470316757

[15] Hauser EB, Venturin GT, Greggio S, Manica E, Zimmer ER, Costa JC. Laplace transform method for ^{11}*C-PIB* two-tissue reversible compartment model with image-derived arterial input function. In: Proceedings of the Ibero Latin American Congress on Computational Methods in Engineering (CILAMCE2017); 5–8 November 2017; Florianopolis. SWGE. 2017. pp. 1-10. DOI: 10.20906/CPS/CILAMCE2017-1077

[16] Hauser EB, Venturin GT, Greggio S, Costa JC. Mathematical modeling to quantify the pharmacokinetic process of [^{18}F]2-fluor-2deoxy-D-glucose (FDG). In: Constanda C, Kirsch A, editors. Integral Methods in Science and Engineering. New York: Birkhäuser; 2015. pp. 301-308. DOI: 10.1007/978-3-319-16727-5-25

[17] Hauser EB, Venturin GT, Greggio S, Borelli WV, Da Costa J. Carotid arterial input function as an inverse problem in kinetic modeling of [^{18}F]2-fluoro-2 deoxy-D-glucose (FDG). Computer Methods in Biomechanics and Biomedical Engineering: Imaging & Visualization. 2019:1-4. DOI: 10.1080/21681163.2019.1647460

[18] Marquardt DW. An algorithm for least-squares estimation of nonlinear parameters. Journal of the Society for Industrial and Applied Mathematics. 1963;**11**:431-441. DOI: 10.1137/0111030

[19] Silva JEM, Furuie SM. Adequacy of compartmental model for positron emission tomography examinations (in Portuguese). Adequação de modelo compartimental para exames de tomografia por emissão de pósitrons. Research on Biomedical Engineering. 2011;**27**(4):231-242. DOI: 10.4322/rbeb.2011.01

[20] Lopresti BJ, Klunk WE, Mathis CA, Hoge JA, Ziolko SK, Lu X, et al. Simplified quantification of Pittsburgh compound B amyloid imaging PET studies: A comparative analysis. Journal of Nuclear Medicine. 2005;**46**(12): 1959-1972

[21] Oliveira F. Logistic Model to Describe the Discrete Curves Activities Obtained from PET Images with [^{18}F] FDG Radiotracer at a Carotid Interest (in Portuguese). Modelo Logístico Para Descrever as Atividades de Curvas Discretas Obtidas de Imagens PET Com Radiofárco [^{18}F]FDG Num Volume de Interesse das Carótídas [Dissertation Lato Sensu]. Porto Alegre: PUCRS; 2019

[22] Cunningham VJ, Jones T. Spectral analysis of dynamic PET studies. Journal of Cerebral Blood Flow Metabolism. 1993;**13**(1):15-23. DOI: 10.1038/jcbfm.1993.5

[23] Khalil MM. Basic Sciences of Nuclear Medicine. Berlin: Springer; 2011. pp. 423. DOI: 10.1007/978-3-540-85962-8

Chapter 6

The Biomechanics of the Anterolateral Ligament

Paul Grimshaw, Christopher Jones, Merilyn Lock and Matthew Longbottom

Abstract

The anterolateral ligament (ALatL) of the knee is an under investigated soft tissue structure of the knee with its existence within the body mostly unknown to the layperson. It was determined that the ALatL has a femoral origin which is either anterior and distal, or posterior and proximal to the origin of the lateral collateral ligament, varying depending on the specimen being investigated. While there have been several studies recently conducted investigating the function of the ALatL of the knee through analysing a number of different factors including origin and insertion along with both physical and mechanical properties, there is still much mystery surrounding this ligament. Hence, further research should be conducted to accurately quantify the importance of the anterolateral ligament to internal tibial rotation stability, and the effect that a damaged anterolateral ligament can have on the stresses experienced by surrounding soft tissue structures of the knee.

Keywords: anterolateral ligament, knee, biomechanics, anterior cruciate ligament

1. Introduction

In late 2013, the world at large was told of a "new ligament" that had been discovered in the human knee, for example by ScienceDaily [1], the New York Times [2] and the British Broadcasting Corporation [3]. The subject of these stories was an anatomical paper by Claes and co-workers [4] that described the anterolateral ligament (ALatL) of the knee.

Looking at the interest generated by this paper does suggest some novelty in this description. At the time of writing, this paper had been cited by 172 other papers on PubMed. **Figure 1** shows the number of papers listed on PubMed, by year, that resulted from a search for the terms "anterolateral ligament" and "knee". The results of this search show a tremendous amount of interest by the scientific community for one small ligament.

Claes and colleagues' study was anatomical, and although it was hypothesised (not tested) that the ALatL is a restraint to internal rotation of the tibia, it could only point to the ligament's "suggested role" in anterior cruciate ligament (ACL) injury-associated instability, highlighting the need for further kinematic evidence.

Two of the reasons for this interest in the anterolateral ligament included (1) a "close association" of anterolateral and anterior cruciate ligament injuries, and (2) residual rotatory instability following anterior cruciate ligament reconstruction [5].

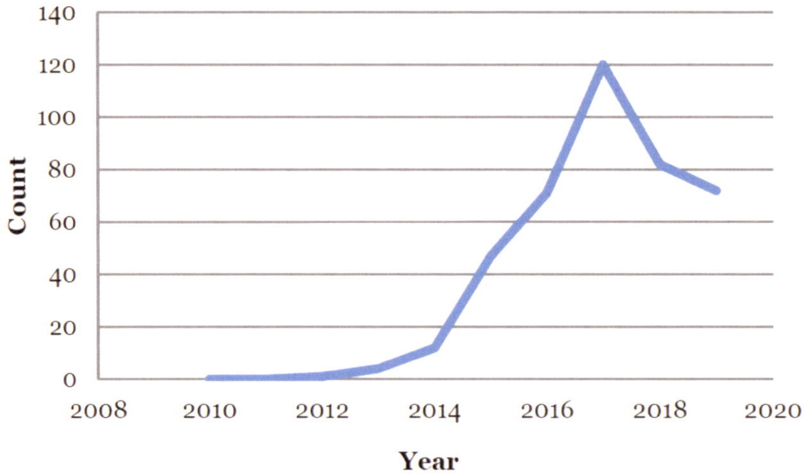

Figure 1.
Number of papers listed on PubMed, by year, that resulted from a search for the terms "anterolateral ligament" and "knee".

It is unclear just how close the association is between the ALatL and ACL injuries [5], although this is not the forum for an in depth analysis. Magnetic Resonance Imaging (MRI) is used to diagnose ALatL injuries, e.g., Claes et al. [6], but MRI analysis has yet to consistently describe the complex anatomy of the lateral side of the knee [7]. Although identification of the ligament itself on MRI appears to be a reliable process, identifying an injury to the ALatL is much less reliable [8].

Thus, it remains to be seen whether it can be established how often the ALatL is injured at the same time as the ACL, or whether an ACL-deficient knee also has an ALatL deficiency.

1.1 Background

The knee of the human body is often considered the most complex joint in the body. Major functions of the knee joint are to provide movement of the lower limb and weight bearing, distributing forces from the femur to tibia. The knee joint provides a range of movements between the femur and tibia through a number of soft tissue structures that articulate motion about the medial and lateral condyles of the two bones. The knee is capable of providing movement in six degrees of freedom with the major movement being flexion/extension with a general range of motion of 0–140°. The joint is most stable when it is in the locked, extended position, however is not mobile. Therefore, this can be the cause of many soft tissue injuries when an external force acts on the knee while it is in this extended position.

Depending on the location, magnitude and direction of the external force applied, damage to different ligaments will occur. This generally occurs when they become taut and overloaded with force. There are essentially eight major ligaments that make up the knee including the anterior cruciate ligament (ACL), posterior cruciate ligament (PCL), medial collateral ligament (MCL), lateral collateral ligament (LCL), patellar ligament, oblique popliteal ligament, arcuate popliteal ligament and the "new" anterolateral ligament (ALatL). All of these ligaments provide stability to the knee depending on their attachment and insertion sites, along with their physical and mechanical properties.

1.2 Aim

This chapter will therefore investigate the properties of the anterolateral ligament of the knee to determine the main functionality it provides. The focus of the chapter will take a biomechanical approach to investigating the anterolateral ligament through the analysis of the physical and mechanical properties. The chapter will conclude with a discussion on its possible injury mechanism and its association with the anterior cruciate ligament.

2. Literature review

Initial research has been conducted to analyse the findings of research papers investigating a number of different features of the ALatL. Firstly, the presence of the ligament is discussed with existing speculation about the existence of the anterolateral ligament in all specimens. Research presenting data about both the physical dimensions of the ligament as well as the mechanical properties will be investigated to determine the functionality of the ligament.

2.1 Presence of the anterolateral ligament

The anterolateral ligament of the knee is a recently defined ligament in the scheme of anatomical history [4], with it only recently being fully recognised as a ligament [9]. There is still confusion surrounding the existence of the ALatL in all specimens with some studies reporting the ligament being found in all subjects, while others have reported discoveries in only as little as 50% of those investigated. The contradiction between studies continues with some demonstrating the ALatL is a capsular structure which has attachment below the lateral meniscus, while others have stated it is extra-capsular, with no such attachment at this site [9].

2.2 Origin and insertion

A study conducted by Claes et al. [4] investigated 41 specimens with an ALatL able to be located in 40 of the 41 samples. In all specimens the major origin of the ligament was located to be fixed to the lateral femoral epicondyle, anterior to the origin of the LCL and proximal and posterior to the insertion of the popliteus tendon. The insertion on the anterolateral tibia was most commonly located midway between Gerdy's tubercle and the fibular head. Within this investigation, all specimens with deficiencies were excluded and knees with a damaged ACL also disregarded [4].

Kennedy et al. [5] conducted an anatomic dissection of 15 non-paired fresh-frozen cadaveric knees to investigate the ALatL. Some major findings from the study included that the ALatL femoral attachment was consistently located posterior and proximal to the attachment of the LCL and the lateral femoral epicondyle. The anterolateral tibial attachment was located approximately midway between the centre of the Gerdy's tubercle and the anterior margin of the fibular head.

A later study conducted by Kosy et al. [9] investigated the presence of the ALatL in 11 specimens to investigate its role within the knee structure. In this study they were able to identify the ALatL in 10 of the 11 specimens. The one specimen, which the ALatL was not able to be identified was also the only specimen which did not have the ACL intact, indicating the ALatL may have been damaged simultaneously in this example. Of the 10 specimens investigated, there was slight variation in the femoral origin of the ALatL, with it found to be posterior and proximal to the LCL attachment in six knees, anterior and distal in three knees, and at the same site in one knee. The tibial attachment was more consistent in location, and was found to be a mean 17.7 (±2.95) mm from Gerdy's tubercle and 12.3 (±3.55) mm from the fibular head.

Helito et al. [10] stated that there is some controversy regarding the femoral attachment site of the ALatL around whether it is anterior and distal or posterior and proximal to the LCL attachment site. Despite this, the tibial attachment site is however consistently defined as being between the Gerdy's tubercle and the fibular head.

In conclusion, studies conducted by Claes et al. [4], Kennedy et al. [5] and Kosy et al. [9] found some discrepancy between the femoral origin of the ALatL and Helito et al. [10] stated that this was a common trait. Claes et al. [4] found the origin to be located anterior and distal to the origin of the LCL, compared to research conducted by Kennedy et al. [5] and Kosy et al. [9] who agreed that the most

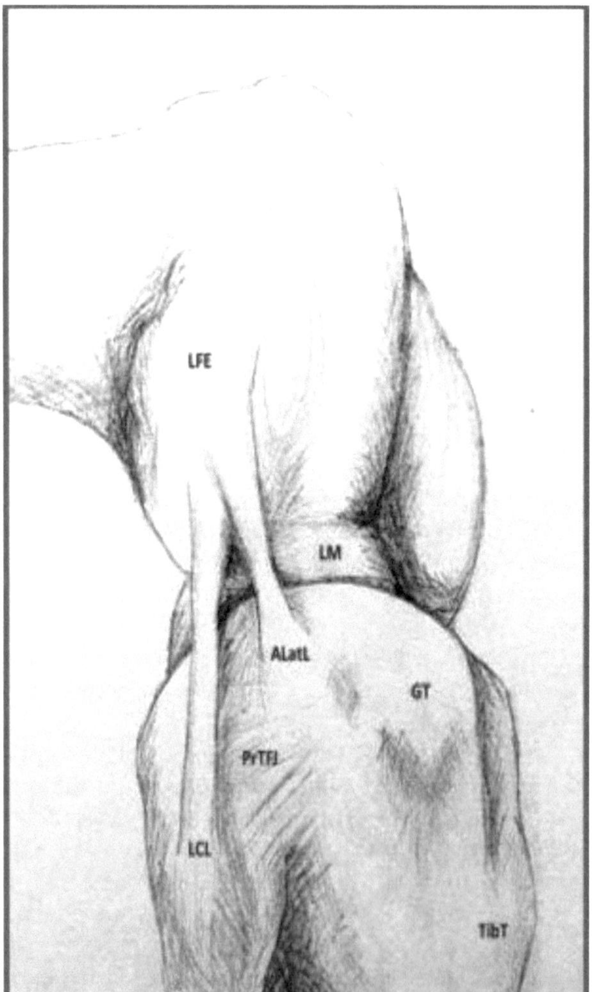

Figure 2.
Sketch of the anterolateral aspect of the knee showing the anterolateral ligament relative to other soft tissue structures. ALatL = anterolateral ligament; GT = Gerdy's tubercle; LCL = lateral collateral ligament; LFE = lateral femoral epicondyle; LM = lateral meniscus; PrTFJ = proximal tibiofibular joint; and TibT = tibial tubercle.

common origin site was posterior and proximal to the LCL attachment. Kosy et al. [9] did however define the origin, in three of the 10 specimens, to be anterior and distal to the ALatL, which agreed with the Claes et al. [4] findings. Hence, it must therefore be concluded that there is variance in the origin of the ALatL in different knee specimens. All four sources however agreed the anterolateral tibial insertion of the ALatL was roughly midway between Gerdy's tubercle and the fibular head [4, 5, 9, 10]. **Figure 2** shows an image that highlights the position of the ALatL on the lateral side of the knee.

2.3 Physical dimensions

Claes et al. [4] found that the measured mean length of the ALatL while in neutral rotation at a flexion angle of 90° was 41.5 (±6.7) mm and in extension was 38.5 (±6.1) mm. This therefore shows that there is some tension in the ALatL during flexion of the knee. Further testing found that the ALatL was mostly in tension while the knee was in a flexed position with an internal rotation applied. The mean width of the ALatLs' investigated by Claes et al. [4], was 6.7 (±3.0) mm and thickness 1.3 (±0.6) mm. Kosy et al. [9] found that of the 10 specimens that the ALatL was able to be located and for a knee flexion angle of 30° and neutral rotation, the mean dimensions were length 40.1 (± 5.53) mm, width 4.63 (±1.39) mm and thickness 0.87 (±0.18) mm. Hence, the values found by Claes et al. [4] and Kosy et al. [9] were similar while also taking into consideration the knee was positioned differently during measurement.

3. Mechanical properties

Zens et al. [11] conducted a biomechanical analysis of the ALatL to determine its mechanical properties allowing for a better understand of its role. Four specimens were investigated in the study. When a load was applied, it was found that all four specimens showed an inter-ligamentous failure at approximately one third of the ALatL's length, distal from the femoral insertion site. The mean ultimate load to failure was 49.90 (±14.62) N, ultimate tension of 32.78 N/mm^2 and the mean ultimate extension distance of 11.89 (±1.56) mm, hence, resulting in a mean extensional stiffness of 2.60 (±0.93) N.

Results from a study conducted in 2015 discovered that the average maximum load that the ALatL was able to handle during a pull-to-failure test was 175 N with a mean stiffness of 20 N/mm [5]. Failure of the ligament occurred through several mechanisms including ligamentous tear at the femoral or tibial origin, mid-substance tear and bony avulsion of the tibial attachment.

Helito et al. [10] also investigated the strength and stiffness of the ALatL of the knee through completing a biomechanical study. The methodology of testing involved 14 knee specimens of which the ALatL was tested for its tensile strength. Throughout testing the strength at the maximum resistance limit, deformation and stiffness were all measured. The mean maximum strength of the ALatL was found to be 204.8 (±114.9) N. The stiffness was 41.9 (±25.7) N/mm and the deformation of the ALatL was 10.3 (±3.5) mm [10]. **Table 1** shows a summary of the findings from the various studies discussed.

The mean maximum strength of the ALatL varied quite significantly between studies with Zens et al. [11] recording a much smaller strength value compared to other sources. Helito et al. [10] had a large variance in strength values from 89.9–319.7 N outlining how much it can vary between specimens. Zens et al. [11] however detailed that the ACL has been proven to have a higher ultimate tensile strength in

	Max. Strength (N)	Stiffness (N/mm)	Deformation (mm)	Ultimate Tension (N/mm^2)
Zens et al. [11]	49.90 (±14.62)	—	11.89 (±1.56)	32.78 N/mm^2
Kennedy et al. [5]	175	20	—	—
Helito et al. [10]	204.8 (±114.9)	41.9 (±25.7)	10.3 (±3.5)	—

Table 1.
Comparison between mechanical properties measured.

younger patients by a factor of 2.5 times. Considering that the previously presented loads of failure are based on specimens of mean age 86.5 ± 1.7 years, there is the possibility that the ALatL ultimate tension may be around 125 N in younger specimens if the same trend follows. The size of the ALatL is expected to reduce with age, hence the possibility for higher ultimate tension in younger specimens. This is more like the range of maximum strength found by other sources. The stiffness of the ligament was found to also differ between Kennedy et al. [5] and Helito et al. [10]. Zens et al. [11] and Helito et al. [10] agreed on the deformation of the ALatL to be around 10–12 mm. Hence, it can therefore be concluded from these studies that the ALatL has approximately the following mechanical properties: maximum strength of 50–200 N, stiffness of 20–42 N/mm, deformation of 30% of its length and tension of 33 N/mm^2. These values and ranges for the mechanical properties of the ALatL were also supported by Patel and Brophy [12].

Drews et al. [13] conducted an investigation to assess the function of the ALatL. This research tested the response of the ALatL through recording the ligaments activity while completing flexion of the knee and a number of rotations and translations. The testing involved eight specimens with the three conditions of: a normal knee with both ACL and ALatL intact, ACL resected (ACL_{res}) and ACL and ALatL resected ($ALatL_{res}$) to be compared. During the flexion testing under unloaded conditions there was no significant difference in internal rotation between the ACL_{res} and $ALatL_{res}$ specimens. With an internal tibial torque of 1–4 Nm applied, internal rotation significantly increased between 60 and 120° in the $ALatL_{res}$ specimen. Anterior tibial translation was also significantly higher at 30° in the $ALatL_{res}$ specimen. When unloaded there were no ALatL strains, however adding different internal tibial torques led to strain of the ALatL starting at an angle of 60° of flexion with a 1 Nm internal torque applied and 15° flexion with 4 Nm internal torque in the intact ligaments. It was also found that the ACL_{res} specimen also had significantly greater ALatL strains under low flexion angles than the intact specimen [13].

4. Analysis of the anterolateral ligament

4.1 Function

Due to the findings of these insertion and origin sites, Claes et al. [4] concluded that given its anatomical location at the anterolateral edge of the knee, it is hypothesised that the ALatL functions as a stabiliser for internal rotation. Dissections found the ALatL to become tense with forced internal rotation between 30 and 90° of knee flexion, as seen in **Figure 3**, however further kinematic analysis is needed to confirm this hypothesis formally. Therefore, research is needed to establish the function of the ALatL and to determine its role in clinical knee joint injuries [4].

Kosy et al. [9] concluded from their study that given the orientation of fibres in the ALatL and the tightening of this structure during internal rotation of the tibia as pictured in **Figure 4**, the structure plays a role in restraining this abnormal movement. Conclusions from the study of Drews et al. [13] were that the ALatL does not have a function under passive motion as there was no change in tibial rotation when the ALatL was removed from the specimen. When extrinsic loads are applied, the ALatL had a slight stabilising effect against anterior tibial shear at low flexion angles. Therefore, it is suggested that the ALatL is supporting the ACL

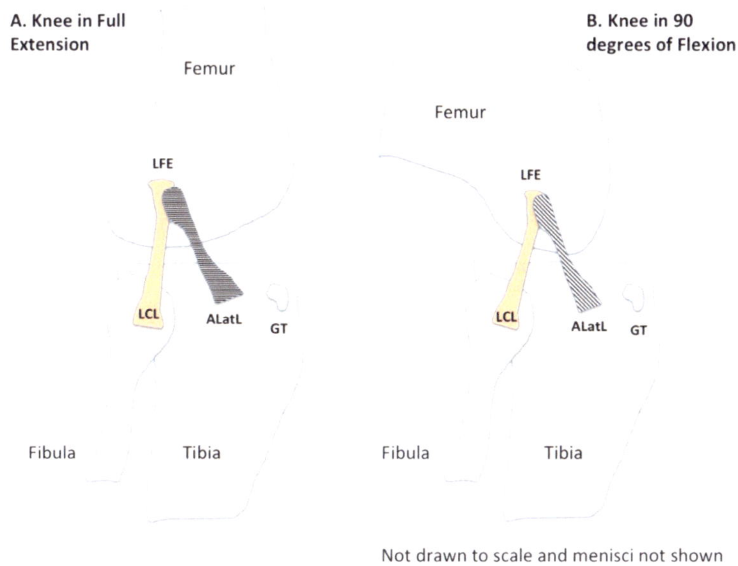

Figure 3.
Anatomical drawing of the anterolateral ligament on the lateral aspect of the human knee. (A) Knee in full extension. (B) Knee in 90° of flexion. ALatL = anterolateral ligament; GT = Gerdy's tubercle; LCL = lateral collateral ligament; and LFE = lateral femoral epicondyle.

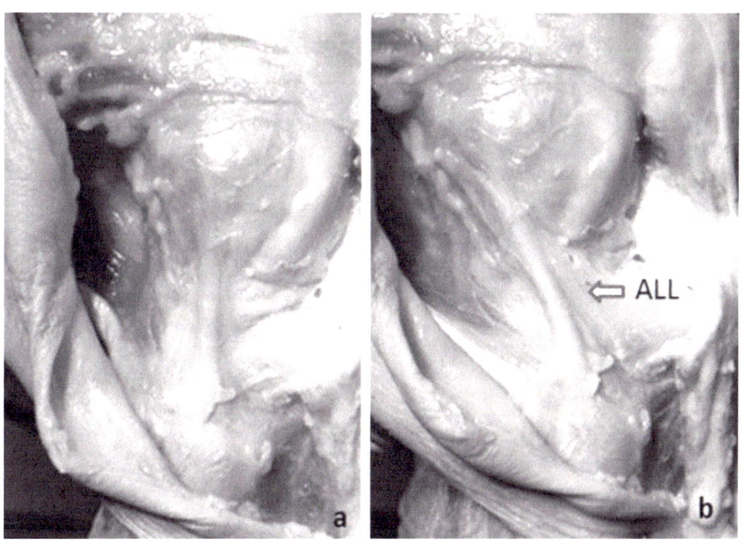

Figure 4.
Photographs demonstrating the tightening of the anterolateral ligament (labelled ALL) between a - a neutral position and b - with internal rotation of the tibia reprinted from, Kosy et al. [9]. Please see the following link for the creative commons licence: http://creativecommons.org/licenses/by/4.0/.

against internal tibial loads to a slight degree. Hence all three sources were in support that the ALatL plays a role in restricting internal tibial rotations [4, 9, 13].

4.2 Kinematics of the anterior cruciate (ACL) and anterolateral (ALatL) ligaments

4.2.1 The anterior cruciate ligament

The largest kinematic difference resulting from ACL deficiency is a substantial increase in anterior tibial translation on application of anterior tibial force [14], which has led to the clinical tests that employ anterior tibial force, for example the Lachman test, where pulling the leg forwards at the knee can result in an increase in translation in the horizontal plane [15].

Surgical replacement of the ACL, or reconstruction, reduces anterior tibial translation to (near-) normal levels, but in some knees there is a rotational instability remaining [16]. This residual instability is thought to lead to the accelerated osteoarthritis seen in the years following ACL rupture and reconstruction [17], and residual rotational instability, diagnosed through the pivot shift test, is associated with poorer outcomes following surgery [18].

In vivo studies of knee joint kinematics following ACL rupture and reconstruction can give a partial view into the ligament's function and its functional loss, but with the potential confounding influence of concurrent injuries to other structures. Further information can be drawn from in vitro laboratory experiments, where damage can be limited to the individual structures. Anterior tibial translation clearly is increased [19–22], but there is contention around rotation in the same plane. Internal tibial rotation as a response to internal tibial torque, defined to occur about a centrally-located axis [23], has been found to be unchanged [24, 25], or slightly increased [22, 26] following ACL ablation, but there has been doubt expressed as to whether this is sufficient to cause clinical problems [22].

The residual rotational instability that is sometimes seen following reconstruction is found during the pivot shift test [27], which involves coupling of internal rotation torque with abduction torque in the frontal plane. Application of these combined loads in the ACL deficient knee leads to normal [19] or slightly increased [28, 29] internal rotation but increased anterior translation [19, 30–32]. This finding has led some researchers to recommend that rotational instability be observed by increased translation [19].

4.2.2 The anterolateral ligament

Injury to the ALatL may cause residual instability in the knee after the ACL has been reconstructed. In a retrospective study of ACL reconstruction patients, those who were diagnosed as also suffering an ALatL tear, by MRI, showed a greater tendency to instability by the pivot shift test following reconstruction [33]. But the picture in relation to the pivot shift test, or in combined internal/valgus torque, is far from clear. What clarity studies of the ALatL provide will therefore be of great interest. Kennedy et al. [34] claimed that "residual rotatory laxity that may be seen clinically following ACL reconstruction may be attributable to an associated anterolateral structure injury". Considering this proposition when looking at ALatL-related kinematics will help to shed light on this issue.

Studies have mostly investigated the ALatL's contribution to kinematics in the ACL-deficient or -reconstructed knee, reflecting the ligament's role as a secondary stabiliser, i.e., having a lesser role than the ACL, and serving that function in the ACL-deficient knee. As knee flexion angle can affect the results of these investigations, the findings of the following studies relate to results found at 30° of flexion.

Application of an 88 N anterior tibial force produced an additional increase in anterior translation after cutting the ALatL in the ACL-deficient knee [35]. This was supported by Tavlo and co-workers [36]. Knees with ACL reconstruction showed no change in anterior translation with an 88 N anterior force after cutting the ALatL, or after reconstructing the ALatL [37]. Thus the role of the ALatL in restraining anterior translation in the presence of either an intact or a reconstructed ACL appears not significant.

ALatL-deficient knees showed an increase in internal rotation on application of 5 Nm internal rotation torque to ACL-deficient knees [35]. Knees with ACL reconstruction showed an increase in internal rotation, compared with the intact state, on internal torque of 5 Nm, and this was increased on cutting the ALatL; rotation in turn decreased to intact levels with ALatL reconstruction [37]. Although Tavlo et al. found that ALatL deficiency led to greater internal rotation in ACL-deficient knees than with ACL deficiency alone, there was no such increase in internal rotation following ACL reconstruction (with the ALatL intact).

A 5 Nm internal rotation torque in combination with a 10 Nm valgus torque increased both anterior translation and internal rotation in ALatL- and ACL-deficient knees, when compared with only ACL-deficiency [35]. On application of a combined loading of 4 Nm internal rotation torque and 8 Nm of valgus torque, sectioning of the ACL increased anterior translation and internal rotation, and sectioning the ALatL increased translation and rotation further [38].

Nitri et al. [37] found that at 30 degrees, anterior translation on combined internal torque (5 Nm) and valgus torque (10 Nm) was not increased after ACL reconstruction, and this was not further changed with ALatL deficiency or reconstruction. In contrast, internal rotation was increased on combination loading in the ACL-reconstructed knee, further increased with ALatL deficiency, and reduced to intact levels following ALatL reconstruction. While loss of the ALatL amplifies the rotational instability of ACL deficiency, where rotational instability has remained after ACL reconstruction, loss/reconstruction of the ALatL increases/alleviates instability.

4.3 Grafts for the anterolateral ligament

Providing a surgeon was to undertake an ALatL reconstruction it would be important to determine the correct method of operation and in particular select the correct replacement graft. Kennedy et al. [5] suggests a graft of the gracilis tendon for an anatomical repair of the ALatL. Mechanical properties of several ligaments as measured by Zens et al. [11] are displayed in **Table 2**. All structures listed are able to provide a sufficient load to failure to replace the ALatL due to the low strength properties of the ligament, however the ultimate tension of the gracilis tendon matches the ALatL the least, with a much higher tension. Based on the data found through the study of Zens et al. [11] other possible graft options, such as the iliotibial band (ITB) or semitendinosis tendon would be a more suitable choice. Kennedy et al. [5] also investigated the best tissue structure to use as a replacement for the ALatL during reconstruction, identifying the semitendinosis tendon and gracilis tendon as appropriate graft selections. This selection was based on their mechanical strength properties and supported choices of both Kennedy et al. [5] and Zens et al. [11].

As well as choosing an appropriate graft for reconstruction of the ALatL, the development of a suitable surgical technique is also important. Kennedy et al. [5] suggested that reconstruction should follow similar techniques as used for other knee reconstructions with grafts used of similar length and properties.

Structure	Ultimate tension [MPa]	Ultimate load to failure [N]	Relation to ALatL (load to failure) [%]
ALatL	**32.78 ± 4.04**	**49.90 ± 14.62**	**100**
ACL	37.80 ± 3.80	1725 ± 269	3457
PCL	35.90 ± 15.20	739–1627	1481
MCL	38.60 ± 4.80	1107 ± 126	2219
Distal sMCL	—	557 ± 55	1116
Proximal sMCL	—	88 ± 36	176
POL	—	256 ± 30	513
Deep MCL	—	101 ± 10	202
LCL	—	309 ± 91	619
PFLC	—	186 N ± 65	373
ITB	19.1 ± 2.9	769 ± 99	1541
Fascia lata	78.7 ± 4.6	628 ± 35	1259
Semitendinosus	88.5 ± 5.0	1216 ± 50	2437
Gracilis	111.5 ± 4.0	838 ± 30	1679

ALatL = anterior lateral ligament; ACL = anterior crutiate ligament; PCL = posterior crutiate ligament; MCL = medial collataeral ligament; sMCL = (superficial) medial collateral ligament; POL = posterior oblique ligament; LCL = lateral collateral ligament; PFLC = popliteofibular ligament complex; ITB = iliotibial band.

Table 2.
Ultimate tension and ultimate load to failure of the ALatL in comparison to other ligaments and possible grafts as researched by Zens et al. [11].

4.4 Discussion

To summarise the results of the ALatL studies described, ALatL-deficiency makes the pattern of ACL deficiency (specifically anterior tibial translation and internal rotation) worse, as there is a greater range of movement with application of a load. In the ACL-reconstructed knee, ALatL-deficiency does not increase anterior tibial translation, but does increase internal rotation, an increase that is reversed after reconstructing the ALatL.

However, the results of ALatL studies cannot simply be superimposed on a background of solid agreement of ACL-related kinematics. While anterior tibial translation on the Lachman test is taken as read, not all researchers have found rotational laxity when applying internal rotation torque, and the pivot shift test may show rotational instability through an increase in translation.

The position and alignment of the ALatL make it logical that it would restrain internal rotation to a degree at least, but that degree aside, it is not always clear what a restraint of internal rotation might be useful for.

Of greater importance in this context is whether instability following ACL reconstruction can be attributed to ALatL injury. Nitri et al. [37] claimed that residual instability on the pivot shift test following ACL reconstruction "reaffirms the theory that tears of secondary restraints... should be properly recognized and treated". Although tears to secondary structures may be responsible for residual instability, it is clear from their results that residual instability results from ACL reconstruction in the presence of an intact ALatL, so at least part of the residual instability is likely to result from the simple fact that ACL reconstruction does not completely restore rotational stability to ACL intact levels. This supports the hypothesis that ACL reconstruction does not completely reproduce ACL-intact kinematics, but given an ALatL lesion its reconstruction could be expected to reduce post-ACL reconstruction instability.

Such a conclusion though is theoretically more difficult to support under in vivo conditions. Less than complete restoration of normal stability appears to be typical following ACL reconstruction, and even with a reliable method of diagnosing ALatL injury, the confounding factor of imperfect ACL reconstruction would still cast uncertainty of the role of the ALatL and the clinical wisdom of reconstructing it.

Within these studies, ALatL reconstruction reverses the rotational deficits seen after ACL reconstruction when the ALatL is deficient. However, in the wider world of ACL research, where the pattern of rotational instability is controversial, and where translation may be a bigger problem, the benefit of recognising and repairing anterolateral ligament injuries is less clear.

5. Summary

The anterolateral ligament (ALatL) of the knee is an under investigated soft tissue structure with its existence within the body mostly unknown to the layperson. It was determined the ALatL has a femoral origin which is either anterior and distal, or posterior and proximal to the origin of the lateral collateral ligament, varying depending on the specimen being investigated. It is not clear exactly why there is such variation. The tibial attachment site was commonly found as midway between Gerdy's tubercle and the fibular head. These origin and attachment sites result in the ALatL running from the lateral femur to the anterior tibia with average physical properties of the ALatL; length of 40–42 mm, width 4–7 mm and thickness 0.8–1.3 mm. The ALatL was found to have mechanical properties on average of; maximum strength of 50–200 N, stiffness of 20–42 N/mm, deformation of 30% of its length and tension of 33 N/mm^2. Studies conducted found that the ALatL is in tension during internal rotations of the knee and therefore acts a stabiliser during internal tibial rotation. All studies presented the issue of lateral instability remaining after patients have undergone isolated ACL reconstruction and had the common belief that this is due to damage caused to lateral soft tissue structures on the knee.

Findings of many studies have not been conclusive enough to confidently suggest ALatL reconstruction as a viable option due to the lower maximum mechanical properties compared to other structures of the knee.

While there have been several studies recently conducted investigating the function of the ALatL of the knee through analysing a number of different factors including origin and insertion along with both physical and mechanical properties, there is still much mystery surrounding this ligament. Hence, further research should be conducted to accurately quantify the importance of the anterolateral ligament to internal tibial rotation stability, and the effect that a damaged ALatL can have on the stresses experienced by surrounding soft tissue structures of the knee.

Author details

Paul Grimshaw[1]*, Christopher Jones[2], Merilyn Lock[3] and Matthew Longbottom[1]

1 School of Mechanical Engineering, The University of Adelaide, Australia

2 Independent Researcher, Australia

3 UniSA Allied Health and Human Performance, School of Health Sciences, The University of South Australia, Australia

*Address all correspondence to: paul.grimshaw@adelaide.edu.au

References

[1] Leuven KU. Surgeons describe new ligament in the human knee. Science Daily. 05 November 2013

[2] Reynolds G. Doctors identify a new knee ligament. New York Times. 13 November 2013

[3] Mundasad S. New ligament discovered in knee, Belgian surgeons say. BBC News. 2013

[4] Claes S, Vereecke E, Maes M, Victor J, Verdonk P, Bellemans J. Anatomy of the anterolateral ligament of the knee. Journal of Anatomy. 2013;**223**(4):321-328

[5] Kennedy MI, Claes S, Fuso FA, Williams BT, Goldsmith MT, Turnbull TL, et al. The anterolateral ligament: An anatomic, radiographic, and biomechanical analysis. The American Journal of Sports Medicine. 2015;**43**(7):1606-1615

[6] Claes S, Luyckx T, Vereecke E, Bellemans J. The Segond fracture: A bony injury of the anterolateral ligament of the knee. Arthroscopy. 2014;**30**(11):1475-1482

[7] Getgood A. Editorial commentary: Magnetic resonance imaging of the knee anterolateral ligament: Does it really matter? Arthroscopy. 2019;**35**(5):1614-1617

[8] Young BL, Ruder JA, Trofa DP, Fleischli JE. Visualization of concurrent anterolateral and anterior cruciate ligament injury on magnetic resonance imaging. Arthroscopy. 2020;(36):1086-1091

[9] Kosy JD, Soni A, Venkatesh R, Mandalia VI. The anterolateral ligament of the knee: Unwrapping the enigma. Anatomical study and comparison to previous reports. Journal of Orthopaedics and Traumatology. 2016;**17**(4):303-308

[10] Helito CP, Bonadio MB, Rozas JS, Wey JMP, Pereira CAM, Cardoso TP, et al. Biomechanical study of strength and stiffness of the knee anterolateral ligament. BMC Musculoskeletal Disorders. 2016;**17**(1):193

[11] Zens M, Feucht MJ, Ruhhammer J, Bernstein A, Mayr HO, Südkamp NP, et al. Mechanical tensile properties of the anterolateral ligament. Journal of Experimental Orthopaedics. 2015;**2**(1):7

[12] Patel RM, Brophy RH. Anterolateral ligament of the knee: Anatomy, function, imaging, and treatment. The American Journal of Sports Medicine. 2018;**46**(1):217-223

[13] Drews BH, Kessler O, Franz W, Dürselen L, Freutel M. Function and strain of the anterolateral ligament. Part I: Biomechanical analysis. Knee Surgery, Sports Traumatology, Arthroscopy. 2017;**25**(4):1132-1139

[14] Butler DL, Noyes FR, Grood ES. Ligamentous restraints to anterior-posterior drawer in the human knee. A biomechanical study. Journal of Bone and Joint Surgery (American Volume). 1980;**62**(2):259-270

[15] Torg JS, Conrad W, Kalen V. Clinical diagnosis of anterior cruciate ligament instability in the athlete. The American Journal of Sports Medicine. 1976;**4**(2):84-93

[16] Tashiro Y, Okazaki K, Miura H, Matsuda S, Yasunaga T, Hashizume M, et al. Quantitative assessment of rotatory instability after anterior cruciate ligament reconstruction. The American Journal of Sports Medicine. 2009;**37**(5):909-916

[17] Simon D, Mascarenhas R, Saltzman BM, Rollins M, Bach BR Jr, MacDonald P. The relationship between anterior cruciate ligament injury and

osteoarthritis of the knee. Advances in Orthopedics. 2015;**2015**:928301

[18] Ayeni OR, Chahal M, Tran MN, Sprague S. Pivot shift as an outcome measure for ACL reconstruction: A systematic review. Knee Surgery, Sports Traumatology, Arthroscopy. 2012;**20**(4):767-777

[19] Diermann N, Schumacher T, Schanz S, Raschke MJ, Petersen W, Zantop T. Rotational instability of the knee: Internal tibial rotation under a simulated pivot shift test. Archives of Orthopaedic and Trauma Surgery. 2009;**129**(3):353-358

[20] Fukubayashi T, Torzilli PA, Sherman MF, Warren RF. An in vitro biomechanical evaluation of anterior-posterior motion of the knee. Tibial displacement, rotation, and torque. Journal of Bone and Joint Surgery (American Volume). 1982;**64**(2):258-264

[21] Oh Y, Kreinbrink J, Ashton-Miller J, Wojtys E. Effect of ACL transection on internal tibial rotation in an in vitro simulated pivot landing. Journal of Bone and Joint Surgery (American Volume). 2011;**93**(4):372-380

[22] Wroble RR, Grood ES, Cummings JS, Henderson JM, Noyes FR. The role of the lateral extraarticular restraints in the anterior cruciate ligament-deficient knee. The American Journal of Sports Medicine. 1993;**21**(2):257-263

[23] Noyes FR, Grood ES, Torzilli PA. Current concepts review. The definitions of terms for motion and position of the knee and injuries of the ligaments. American Journal of Bone and Joint Surgery. 1989;**71**(3):465-472

[24] Lane JG, Irby SE, Kaufman K, Rangger C, Daniel DM. The anterior cruciate ligament in controlling axial rotation: An evaluation of its effect. The American Journal of Sports Medicine. 1994;**22**(2):289-293

[25] Wünschel M, Müller O, Lo J, Obloh C, Wülker N. The anterior cruciate ligament provides resistance to externally applied anterior tibial force but not to internal rotational torque during simulated weight-bearing flexion. Arthroscopy. 2010;**26**(11):1520-1527

[26] Hemmerich A, van der Merwe W, Batterham M, Vaughan CL. Knee rotational laxity in a randomized comparison of single- versus double-bundle anterior cruciate ligament reconstruction. The American Journal of Sports Medicine. 2011;**39**(1):48-56

[27] Galway H, Macintosh D. The lateral pivot shift: A symptom and sign of anterior cruciate ligament insufficiency. Clinical Orthopaedics and Related Research. 1980;**147**:45-50

[28] Kondo E, Merican AM, Yasuda K, Amis AA. Biomechanical comparison of anatomic double-bundle, anatomic single-bundle, and nonanatomic single-bundle anterior cruciate ligament reconstructions. The American Journal of Sports Medicine. 2011;**39**(2):279-288

[29] Yamamoto Y, Hsu W-H, Woo SLY, Van Scyoc AH, Takakura Y, Debski RE. Knee stability and graft function after anterior cruciate ligament reconstruction: A comparison of a lateral and an anatomical femoral tunnel placement. The American Journal of Sports Medicine. 2004;**32**(8):1825-1832

[30] Woo SLY, Kanamori A, Zeminski J, Yagi M, Papageorgiou C, Fu FH. The effectiveness of reconstruction of the anterior cruciate ligament with hamstrings and patellar tendon : A cadaveric study comparing anterior tibial and rotational loads. Journal of Bone and Joint Surgery (American Volume). 2002;**84**(6):907-914

[31] Zantop T, Herbort M, Raschke MJ, Fu FH, Petersen W. The role of the anteromedial and posterolateral bundles of the anterior cruciate ligament in anterior tibial translation and internal rotation. The American Journal of Sports Medicine. 2007;**35**(2):223-227

[32] Kanamori A, Woo SLY, Ma CB, Zeminski J, Rudy TW, Li G, et al. The forces in the anterior cruciate ligament and knee kinematics during a simulated pivot shift test: A human cadaveric study using robotic technology. Arthroscopy. 2000;**16**(6):633-639

[33] Castelli A, Zanon G, Jannelli E, Ivone A, Ferranti Calderoni E, Combi A, et al. The role of the anterolateral ligament in knee's biomechanics: A case–control retrospective study. European Journal of Orthopaedic Surgery and Traumatology. 2019;**2019**:1-6

[34] Kennedy MI, LaPrade CM, Geeslin AG, LaPrade RF. An overview of clinically relevant biomechanics of the anterolateral structures of the knee. Techniques in Orthopaedics. 2018;**33**(4):213-218

[35] Rasmussen MT, Nitri M, Williams BT, Moulton SG, Cruz RS, Dornan GJ, et al. An in vitro robotic assessment of the anterolateral ligament, Part 1: Secondary role of the anterolateral ligament in the setting of an anterior cruciate ligament injury. The American Journal of Sports Medicine. 2016;**44**(3):585-592

[36] Tavlo M, Eljaja S, Jensen JT, Siersma VD, Krogsgaard MR. The role of the anterolateral ligament in ACL insufficient and reconstructed knees on rotatory stability: A biomechanical study on human cadavers. Scandinavian Journal of Medicine & Science in Sports. 2016;**26**(8):960-966

[37] Nitri M, Rasmussen MT, Williams BT, Moulton SG, Cruz RS, Dornan GJ, et al. An in vitro robotic assessment of the anterolateral ligament, Part 2: Anterolateral ligament reconstruction combined with anterior cruciate ligament reconstruction. The American Journal of Sports Medicine. 2016;**44**(3):593-601

[38] Thein R, Boorman-Padgett J, Stone K, Wickiewicz T, Imhauser C, Pearle A. Biomechanical assessment of the anterolateral ligament of the knee: A secondary restraint in simulated tests of the pivot shift and of anterior stability. Journal of Bone and Joint Surgery (American Volume). 2016;**98**(11):937-943

Chapter 7

Clinical and Experimental Biomechanical Studies Regarding Innovative Implants in Traumatology

Paul-Dan Sirbu, Wilhelm Friedl, Dan Mihailescu, Liliana Savin, Andrei Scripcaru, Norin Forna, Mihnea Theodor Sirbu, Mihaela Pertea and Razvan Cosmin Tudor

Abstract

Fracture treatment has experienced a fascinating evolution in the last years. The aim of this chapter is to reveal some clinical and biomechanical studies regarding innovative implants. After a short introduction (1), we intend to present our results regarding (2) dynamic condylar screw versus condylar blade plate in complex supracondylar femoral fractures; (3) biomechanical analysis of four types of implants in humeral fractures; (4) clinical and experimental studies for optimal stabilization of trochanteric fractures: the gliding nail; (5) intramedullary XS nail for pilon and ankle fractures: design, biomechanics, and clinical results; (6) the XS nail for the treatment of patella and olecranon fractures; and (7) plates with polyaxial stability for fractures of distal radius and proximal humerus. In conclusion, the authors highlight the advantages of these innovative implants in difficult trauma cases.

Keywords: fracture treatment, biomechanical, gliding nail, XS nail, polyaxial stability

1. Introduction

Hardship resulted from difficult periarticular fractures and numerous complications following the usage of classic plates and open reduction techniques and determined the development of innovative implants and new types of surgical techniques. The combination of three imperative criteria for fracture treatment (high biomechanical stability, anatomical reduction of the articular surface, and percutaneous insertion with minimal soft tissue damage) has led to the development of a new generation of implants.

The aim of this chapter is to present the advantages of innovative implants in traumatology: dynamic condylar screw (DCS) versus condylar blade plate (CBP) in distal femoral fractures; the optimal implant in humeral fractures; XS nail in ankle, patella, and olecranon fractures; the gliding nail for trochanteric fractures; and plates with polyaxial stability.

2. DCS versus CBP in complex supracondylar femoral fractures: a biomechanical study

Distal femoral fractures represent a challenge for orthopedic surgeons, and despite numerous biomechanical studies, the optimal implant is still controversial [1–4]. However, while plates with angular stability and retrograde interlocking nails are nowadays the best choice for treatment, CBP and DCS were the most used implants until the development of these innovative implants [5].

In a biomechanical study from 2009, the authors compared the mechanical rigidity of the bone/implant (DCS or CBP) construct in complex supracondylar femoral fractures [5, 6].

Twelve synthetic composite femoral bones were fixed in the distal part with six DCS and six CBP, and then, the authors performed by osteotomy a bone defect of 1.5 cm to simulate a complex supracondylar fracture type A3/AO (**Figure 1**).

The femurs were sectioned in the midshaft, and the proximal part of the distal fragments was fixed in a metallic adapter sleeve. The bone-implant constructs were tested for seven types of loading: (1) internal compression; (2) external compression; (3) anterior compression; (4) posterior compression; (5) axial compression; (6) external torsion; and (7) internal torsion.

The compression tests were realized up to 350 N, and the applied torsion attended 25 Nm. The tests were repeated six times in order to establish the statistic dispersion. All the measurements for DCS were realized with or without compaction screw.

The compression force and loading force were measured by a M221B04 (PCB Piezotronics force transducer), while linear deformation values for the compression were measured using two inductive transducers applied in frontal axis (TD1) and sagittal axis (TD2) (**Figure 2**).

Data acquisition was realized by a six-channel admittance bridge, an interface board, and a digital data acquisition system DAQ1200 connected to a laptop.

According to study measurements, by reporting the loading/unloading force to the transducer (TD1 and TD2) displacement, we represented hysteresis cycles as diagrams for the femur/DCS (with and without compaction screw) and for femur/CBP (**Figure 2**).

By analyzing these measurements and diagrams, the authors obtained preliminary results regarding DCS versus CBP, which were statistically processed by

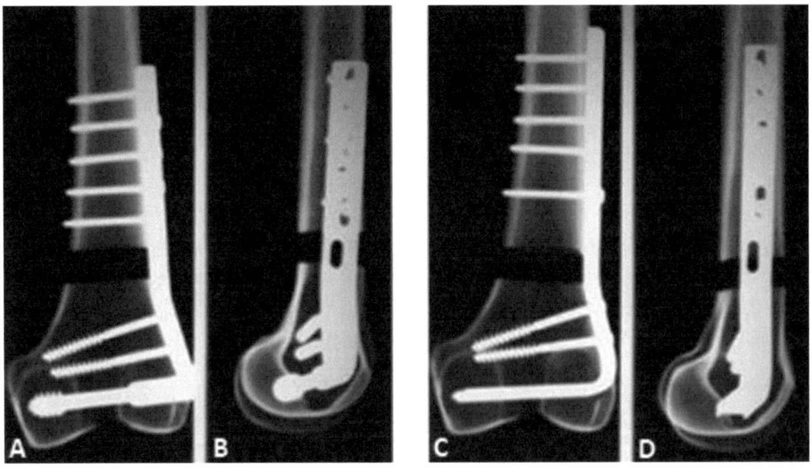

Figure 1.
(A–B) Radiographic aspect of a synthetic composite femur with osteotomy and fixed with DCS and (C–D) radiographic aspect of a synthetic composite femur with osteotomy and fixed with CBP.

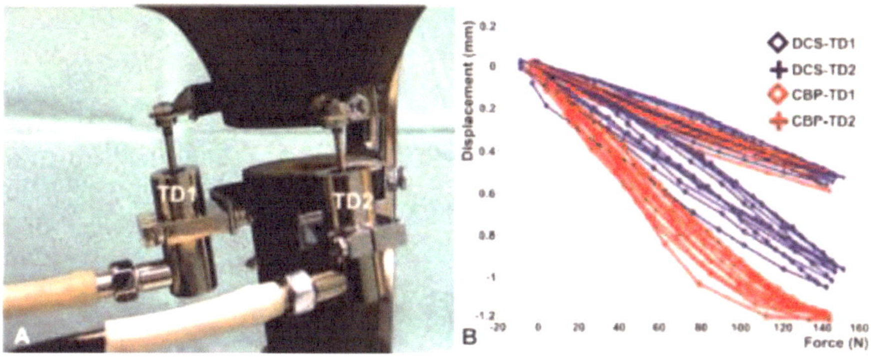

Figure 2.
(A) Deformation measuring methods. Transducers: TD1—frontal axis; TD2—sagittal axis; (B) internal compression (DCS/CBP). Six loading tests. TD1 deformations, 12–16% higher for CBP than DCS; TD2 deformations, comparable for CBP versus DCS. Negative values (osteotomy closure). Mechanical hysteresis (both implants).

calculating the mean stiffness (square mean error) and the "p" value. The stiffness of the two implants differs significantly if $p < 0.05$ (95% reliability).

After the interpretation of the statistical study (DCS with compaction screw/CBP), the authors noticed that the femur-DCS construct is more stable in all compression types except the posterior and axial one, where CBP seems to be more resistant for TD2 transducer.

While, in 2002, Jaakkola et al. [5] found that there is no biomechanical advantage of CBP over DCS on plastic bones, this biomechanical study on synthetic composite femurs suggests that DCS is better than CBP in most loading tests, and the compaction screw for DCS confers an increased stability to the construct.

3. Biomechanical analysis of four different types of implants in humeral shaft fractures

The surgical treatment for humeral shaft fractures is still debatable as long as, according to comparable rate of union, the "nailers" sustain a close intramedullary technique (despite an increased risk of shoulder pain), while the "platers" emphasize the advantages of the open reduction internal fixation (ORIF; with no shoulder morbidity, despite the risk of radial nerve injury) [7, 8]. Some studies in the literature advocate the mechanical advantages of intramedullary nails [9], while other authors enhance the advantages of weight bearing on crutches with plate fixation for patients with associated lower limb fractures [10].

The aim of a biomechanical study from 2010 was to evaluate the mechanical behavior of four different types of implants used for internal fixation of comminuted humeral shaft fractures [11].

In 12 synthetic composite bones, the authors simulated a comminution in the middle third of diaphysis by removing a 38-mm thick fragment. The bones were separated in four groups, and the fractures were instrumented with four types of implants: (1) a locked compression plate (LCP; Synthes®) with six holes; (2) an intramedullary static locked (Medimetal®) nail inserted in a retrograde manner; (3) a long monoaxial locked plate type AxSOS (Stryker®) fixed with four screws (with a longer "working length"); and (4) a classic 13 holes long dynamic compression plate (DCP) with six cortical screws (**Figure 3**).

The mechanical tests were performed on a loading machine LLOYD LRX 5kN (UK), which allows traction-compression tests with forces up to 5000 N, on

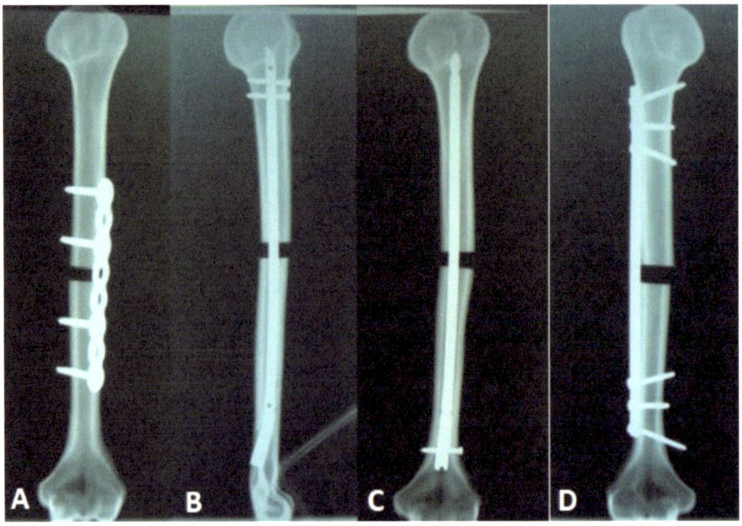

Figure 3.
(A) Locked plate; (B and C) locked nail; and (D) DCP-buttress plate.

variable speeds (0.01–800 mm/min) and an accuracy of minimum 0.2%. The compression forces were measured using the force cell of the machine (0.01% precision), and the deformations were measured with a resolution of 0.1 microns. For the testing trials, we used the Nexygen and Ondio producer provided software. All of the constructs were submitted to torsion essays in external and internal rotation as to obtain the same amount of torque [11].

According to the measured values, the authors obtained load-deformation diagrams corresponding to the four types of implants and two types of torsion loading (**Figure 4**).

The load-deformation diagrams were compared and statistically analyzed for each type of implant.

The shorter LCP proved to be the most rigid implant for each type of loading essay, the mean values of the loading being the highest in the entire group. This construct with a short angular stable plate and a small working length is unfortunately a stiff device that concentrates stress at the bone-screw interface.

The intramedullary locked nail showed to be the most elastic implant of all types of loading but, at the same time, the less rigid implant in torsion.

The classic DCP demonstrated, surprisingly, in all types of torsional loading, a mechanical behavior close to the AxSOS angular stable plate; this result is related to the fact that by using longer plates with few screws placed far from fracture site

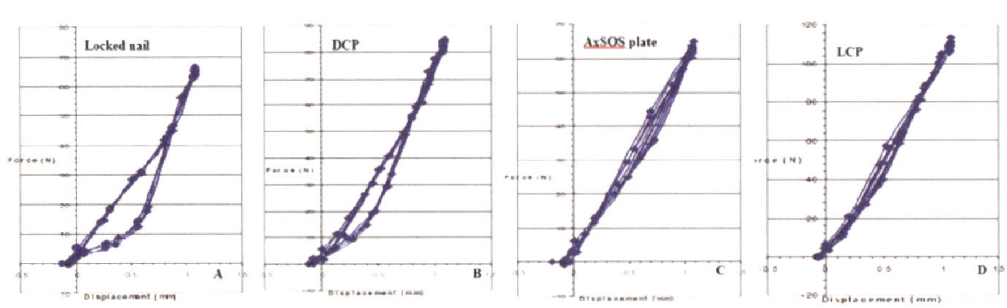

Figure 4.
Load-deformation diagrams corresponding to the four types of metallic implants loaded in external rotation: (a) locked nail; (B) DCP; (C) AxSOS plate; and (D) LCP.

(bigger working length), the torsional stress is distributed more evenly on the entire length of the plate, the mechanical stress between bone and screws is reduced, and the entire construct became an elastic one.

4. Clinical and experimental studies for optimal stabilization of trochanteric fractures: the gliding nail

The high incidence of osteoporosis in the elderly population and the high mechanical load on the proximal femur make the trochanteric region a common fracture site.

Due to the different types of fracture patterns, each with its own characteristics, a universally applicable implant is very difficult to set. The fixation strength for a pertrochanteric fracture is determined by different variables such as bone quality, bone fragment geometry, fracture reduction, implant design, and implant placement [12].

Numerous studies show that the implant used, as well as its placement, is very important for a successful outcome [13–15].

Depending on the implant position, the types of implants used can be extramedullary or intramedullary.

The dynamic hip screw (DHS) and the blade plate are commonly used implants in pertrochanteric fractures. Due to the longer length of the lever arm, they are subjected to a higher bending stress, making the risk of fatigue fractures or cutout higher than intramedullary implants (**Figure 5**). Moreover, the placement of such an implant requires large incisions with soft tissue damage and deperiostation. In these conditions, the local vascularization is greatly impaired, and the risk of local complications is higher.

Furthermore, immediate restoration of weight bearing is not entirely possible, and considering the mean age of the patients, this is of vital importance.

Intramedullary implants existed since the development of the Y-profile Küntscher nail and due to the implant position in the medullary canal, they all share a less bending force compared to extramedullary implants. Also, the surgical technique required for their implantation minimizes the soft tissue damage [16].

The most common intramedullary implants are the gamma nail and the proximal femoral nail (PFN). Since 1994, extensive clinical and experimental investigations conducted in Germany have led to the development of an intramedullary gliding nail (GN). This system has the biomechanical advantages of an intramedullary locked implant, and because of the double-T angle blade profile, the gliding screw system creates an increased resistance [17] (**Figure 6**).

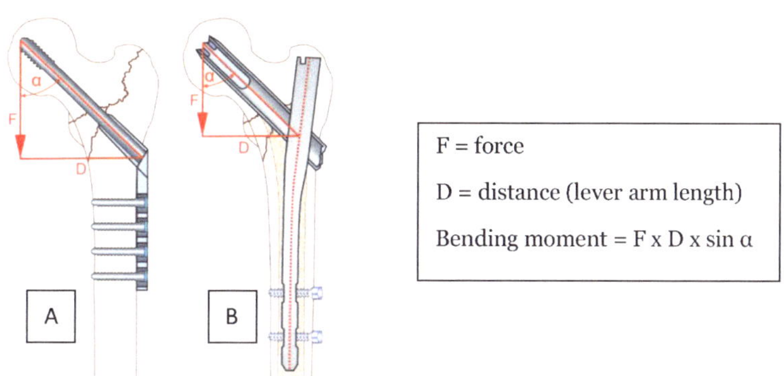

Figure 5.
(A) Extramedullary DHS system and (B) intramedullary GN system.

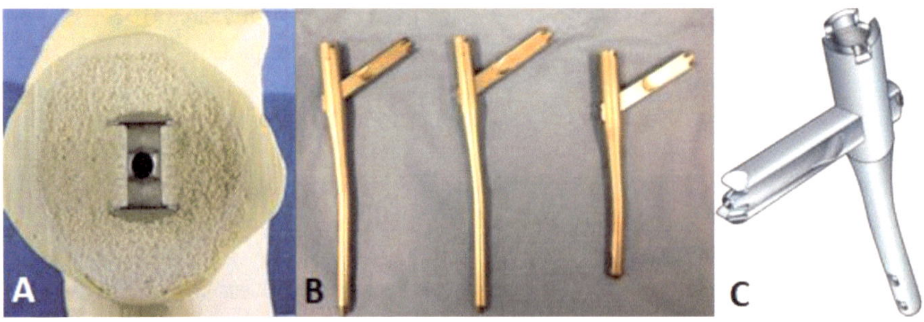

Figure 6.
(A) Axial view of the double-T blade profile inside the femoral head (sawbone); (B) lateral view of the GN in three sizes; and (C) 3D view of the gliding nail.

The double-T profile has a higher stiffness due to its rotational stability and due to its reduced risk of damage in osteoporotic bone. The nail curvature of 6° in frontal plane and straight in sagittal plane allows the entry point on the tip of the trochanter (thus having a lower risk of circulatory disorders to the femoral head than opening the piriformis fossa) (**Figure 7**).

Another important characteristic is the dynamic impaction possibility in the femoral neck direction with dynamic stability in the femoral shaft direction (**Figure 8**).

The first study from 1996 [18], which compared the gliding nail system with the gamma nail, showed better intraoperative and postoperative results for the GN. The rate of intraoperative complications for the GN was 2.7%, while for the gamma nail, it was between 17.2 and 42.2%. The difference in outcome is highlighted by the long-term results, where the gliding nail had only 3.9% rate of complications, while the gamma nail had 6–13.8% [18].

Following the promising results, a biomechanical study from 1998 showed the importance of the blade geometry for the stability of fixation in proximal femoral fractures. The alternating load examinations on Sawbone femoral heads revealed no instability of the implant after 100,000 cycles at a load of 2000 N. The displacement of the double-T blade after 1000 cycles at 1500 N was 1–4 mm, while for the 10 mm, screw of the gamma nail was 4–8 mm [19].

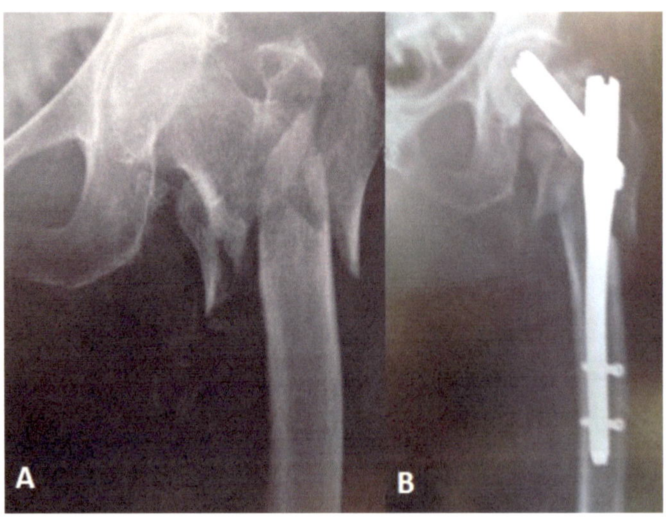

Figure 7.
(A) AO/OTA A3.3 pertrochanteric fracture and (B) osteosynthesis with gliding nail.

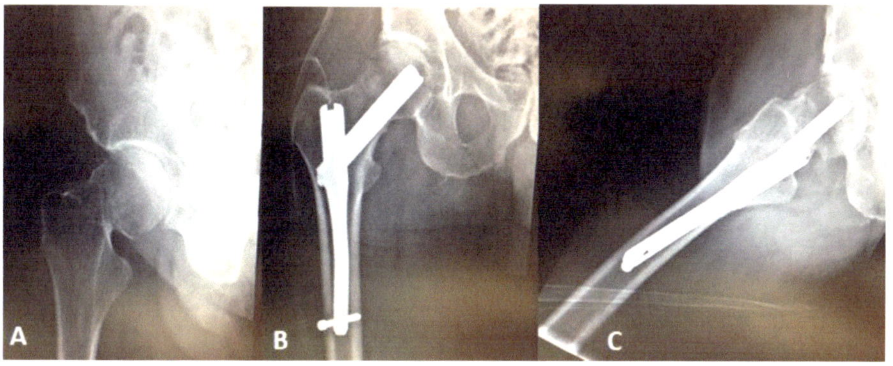

Figure 8.
(A) Garden IV femoral neck fracture and (B and C) closed reduction and osteosynthesis with gliding nail.

In 1998, Friedl et al. published a study on 186 patients with pertrochanteric fractures treated with gliding nail; the authors revealed a very good outcome with low complication rate, especially for intraoperative complications (1.1%) and late local complications (4.9%) [20].

The superiority of the gliding nail system over the DHS or gamma nail was highlighted by the authors in a series of clinical and biomechanical studies.

The most recent study published in 2009 carried out over a period of 5 years and studied 501 patients with trochanteric and subtrochanteric fractures operated only with the GN system and immediate weight bearing. The results revealed that local complications of these difficult fractures like cutout or severe impaction can be avoided by using the gliding nail system [21].

5. Locked intramedullary XS nail for ankle and pilon fractures: design, biomechanics, and clinical results

Fractures around the ankle are very frequent injuries, and the aim of the treatment is reconstruction of the anatomy with stable and minimally invasive osteosynthesis techniques [22, 23] while avoiding further trauma of soft tissue in a local region with anatomical peculiarities: the skin is thin, with limited mobility, with almost nonexistent skin excess, a very poorly represented subcutaneous soft tissue, and poor blood supply [24, 25].

Plate osteosynthesis is the "gold standard" procedure for distal fibula, pilon, and lower leg fractures, but Zaghloul et al. reported a rate of 21.5% complications, with 2% infections and 10.8% operative revisions [26].

The severe wound complications associated with an extramedullary implant due to the compromised blood supply (arterial occlusion diseases, diabetes, and post-thrombotic sequels) and the thin soft tissue envelope require removal of the plate (with secondary stability impairment) and additional challenging reconstructive technical solutions including split-thickness skin grafts and local or locoregional flaps [27–29]. With regard to the use of split-thickness skin grafts, they are often impossible to use in the case of soft part defects in the ankle, due to bone or tendinous exposure. Lately, a solution worthy of consideration is the use of negative pressure therapy, so that a good, vascularized bed can be created, which will allow the use of a split-thickness skin graft (**Figure 9**).

Intramedullary implants had biomechanical advantages over plates by reducing the lever arm and increasing the stability of the construct [27, 30]. The limitations of simple wires, intramedullary pins, and distally locked flexible nails,

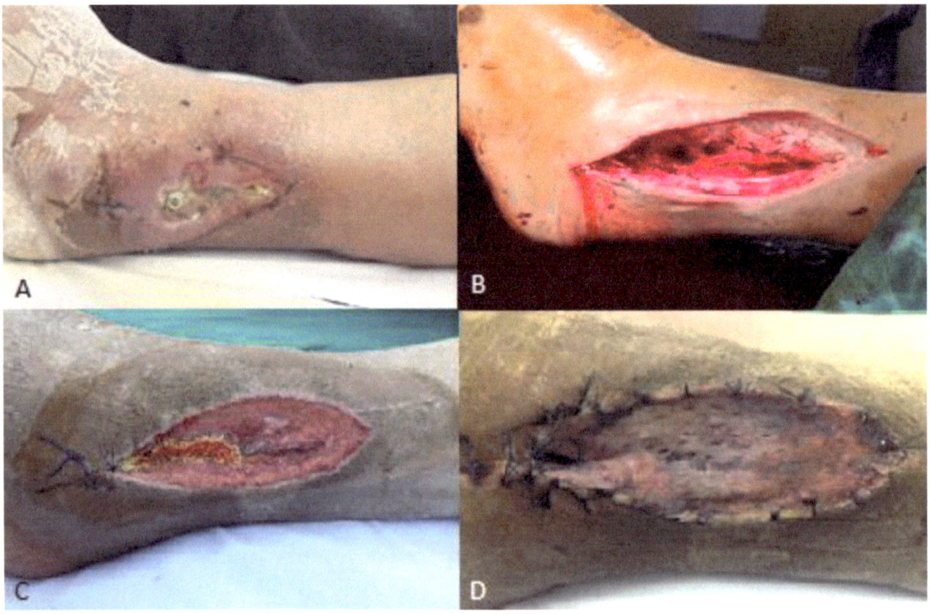

Figure 9.
Wound complication after fibular plate osteosynthesis; (A) necrosis and exposure of the osteosynthesis material; (B) soft tissue defect after debridement; (C) granular bed after negative pressure therapy; and (D) granular bed covered with skin graft.

regarding no maintenance of length and rotation, prompted the surgeons to look for alternative techniques for osteosynthesis; the crucial requirements were minimally invasive approaches with less soft tissue irritation and high biomechanical stability. When a soft tissue problem occurs, the treatment is much easier because there is no implant on the bone surface, so that only excision and split skin graft can be performed.

From 1999, some investigators in Germany [27, 30] started to use a novel intramedullary cannulated small-diameter straight locked nail system for the fibula osteosynthesis, which was primarily designed for the treatment of fractures under tension: the XS nail. The intramedullary device length was improved to the needs of the fibula osteosynthesis (**Figure 10**). The advantages of this nail allowed their use for fractures of the fibula and tibial pilon.

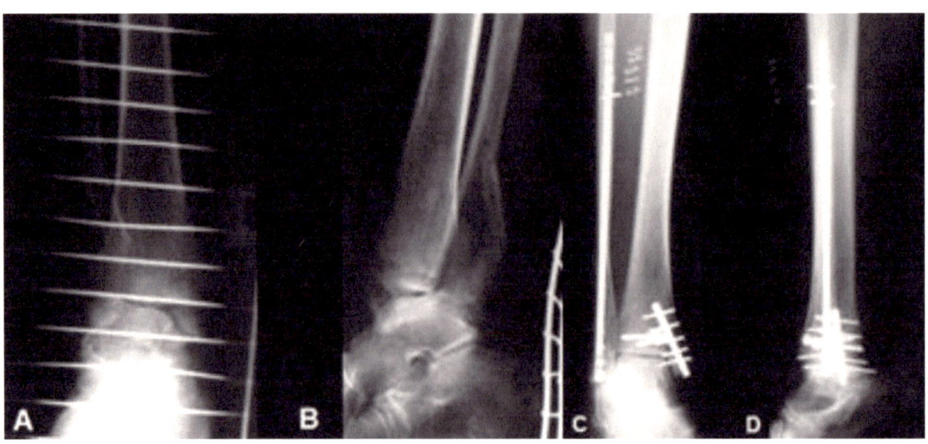

Figure 10.
Ankle fracture before and after XS nail osteosynthesis.

The XS(L) nail used in the treatment of malleolar fractures has a rounded figure and a width varying from 4.5 to 3.5 mm in the case of XXS(L) nail [27, 28, 30]. The type of fracture dictates the number of locking holes needed, with an availability between 4–11 spaced (9 mm) locking holes and 3, 4, 6, 8, or 10 locking holes for the smaller XXS(L) nail (**Figure 11**). Two holes closest to the fixation have a longitudinal oval alignment. For XSL long nails, there are available lengths of up to 272 mm, and 197 mm for the XXSL nails [28].

Before starting with the clinical application of this innovative implant, experimental tests were performed comparing a standard plate and compression screw with the XS nail in fibula osteotomies [27, 31].

Eighteen sawbones were used to create Weber type B, type C, and suprasyndesmotic fractures by resection osteotomy. In each group, three sawbones were "treated" by plate and compression screw, and three with XS nail. Load was applied in 1000 cycles of 40–100 N with a frequency of 1 Hz using a MTS® machine (www.mts.com). Both plastic and total deformations were recorded. The tests showed that the nail group showed a lower deformation and higher stability.

Based on these promising results, from May 2000 to January 2002, the surgeon designers operated 194 ankle joint fractures (one-third bimalleolar injuries) using the innovative XS nail [27, 30]. The results (according to Olerud score) were encouraging (excellent outcome—58.6%, a good one—33.3%, a fair one—5.5%, and an unsatisfactory outcome—2.5%), with very few soft tissue problems and only one nonunion.

The nail insertion is realized after open fracture reduction in displaced ankle fractures (**Figure 12**) and percutaneously for pilon and lower leg fractures (with extraarticular involvement) [27, 28, 30]. The preoperative planning is crucial including the analysis of diameter of the medullary canal up to the optimal level of stabilization. XS nail is not indicated in rare cases when the canal is too narrow, and for a very thin medullary space, the XXS nail should be used. It is essential that the medullary canal must have a diameter of at least 3 mm in case of XS nail osteosynthesis [27, 28, 30]. The nail insertion requires the division of the retinaculum distal to the fibula and retraction of the peroneal tendons.

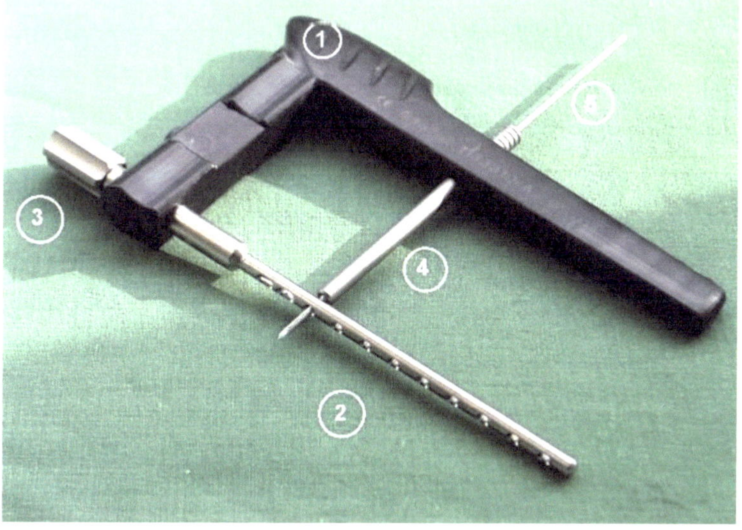

Figure 11.
Standard XS nail with aiming device: (1) X-ray transparent carbon aiming device; (2) XS standard (4–5 mm) nail, here 12-hole nail; (3) adapter for the nail on the aiming device; (4) lateral drill sleeve for the locking wires; and (5) cross-locking threaded K wire.

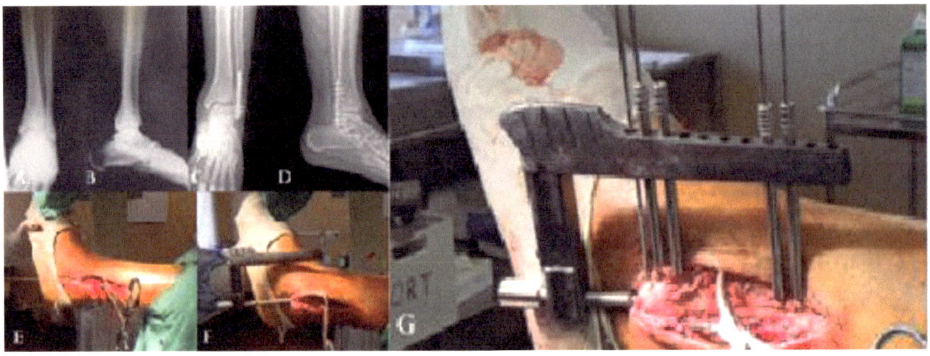

Figure 12.
XS nail osteosynthesis for a left distal fibula fracture; (A and B) preoperative X-rays; (C and D) postoperative X-rays; (E) insertion of the guiding wire; (F) insertion of the nail with the aiming device; and (G) locking with transfixing threaded wires.

After open reduction of distal fibular fracture and compression with a reduction forceps, a guide wire (1.6–2 mm thickness) is inserted under fluoroscopic control in the medullary canal. The position of the wire is verified in two radiological views, and a cannulated drill is used to the desired length of the nail, using the same diameter. The best stability is obtained by placing a nail, which is at least 2.7 cm (3 holes) longer than the fracture site.

A crucial step is to choose the biggest nail diameter fitting into the medullary canal, in order to obtain the stability of the syndesmosis without fibulotibial set screw [27, 28].

The radiolucent aiming device is used to place the nail tightly in the canal, and afterward, cross-locking is realized with threaded K wires at 90°; 2.4 or 2.0 mm wires are used depending on the type of nail (XS or XXS). Maximum stability is obtained by securing the nail with two proximal and two distal wires. The hole diameter is 0.2 mm smaller and ensures an angle stable locking for the wires. Additional interfragmentary compression can be achieved in transverse or short oblique fracture by placing a compression screw after the removal of the aiming device. For oblique or comminuted fracture, the compression realized with reposition forceps can be preserved by threaded wires that cross the fracture site [27, 28, 30]. In type C fractures, with a larger medullar canal than the nail diameter, an additional fibulotibial set screw must be used in order to ensure a stable syndesmosis [28].

At the end of the surgery, the threaded wires are cut with a special device, in order to minimize the implant over the bone surface.

The XS nail can additionally be used in percutaneous technique for concomitant fractures of the fibula in distal metaphyseal fractures of the lower leg, pilon fractures and for fixation of the tibia following joint reconstruction [28, 32, 33]. A perfect anatomical reduction of the fibula is not needed, so no additional damage will compromise the thin soft tissue and impaired blood supply (**Figure 13**).

The percutaneous osteosynthesis with an XS nail is first performed for the fibular fracture and then tibia is fixed with one XS nail introduced from the medial malleolus to the proximal lateral cortical bone; if it is necessary, a second XS nail is inserted from the distal lateral tibia aiming at the proximal medial cortical bone [28, 32, 33].

The minimally invasive osteosynthesis of pilon fracture using the XS nail, as well as the absence of plates on the bone surface, reduces significantly the healing problems and the rate of complication for these difficult fractures [33].

All the clinical and biomechanical studies enhanced that XS and XXS nails are important alternatives to classic plate osteosynthesis for distal fibula, pilon, or

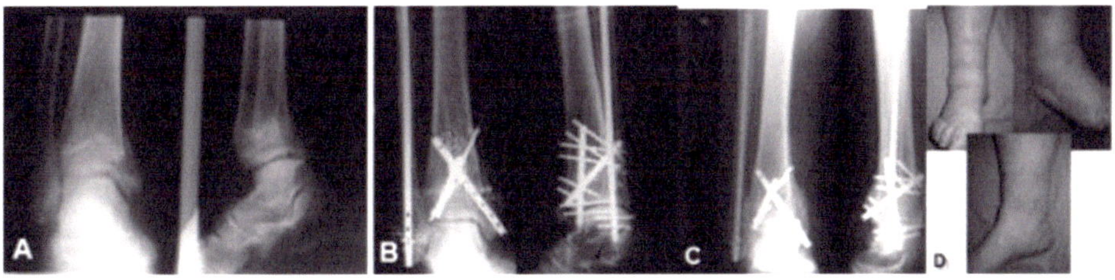

Figure 13.
Distal metaphyseal lower leg fracture and XS nail osteosynthesis.

lower leg fractures while avoiding soft tissue complications and ensuring a stable fracture fixation with a higher weight bearing tolerance.

These advantages highlight the XS nail as an "ideal" implant for patients with poor bone and soft tissue healing problems due to vascular conditions, diabetes, or trauma.

6. The XS nail for the treatment of fractures under tension: patella and olecranon

In patella fractures, the surgical treatment with the AO tension belt osteosynthesis system is the golden standard today, but the results are not always good on the long term. Dislocation and functional deficit (limited mobility) can be as high as 20–50% of all cases [34].

One explanation is that the tendon insertions and the retinaculum create a gap between the tension-band wires and the bone, thus a very tight fixation cannot be achieved; because of this, loosening occurs after loading [35].

Moreover, due to the fact that the AO tension band is placed on the anterior surface of the patella, there is a distraction in the fracture site on the articular surface, which causes fracture gaps, dislocations, nonunion, and finally implant failure (**Figure 14**).

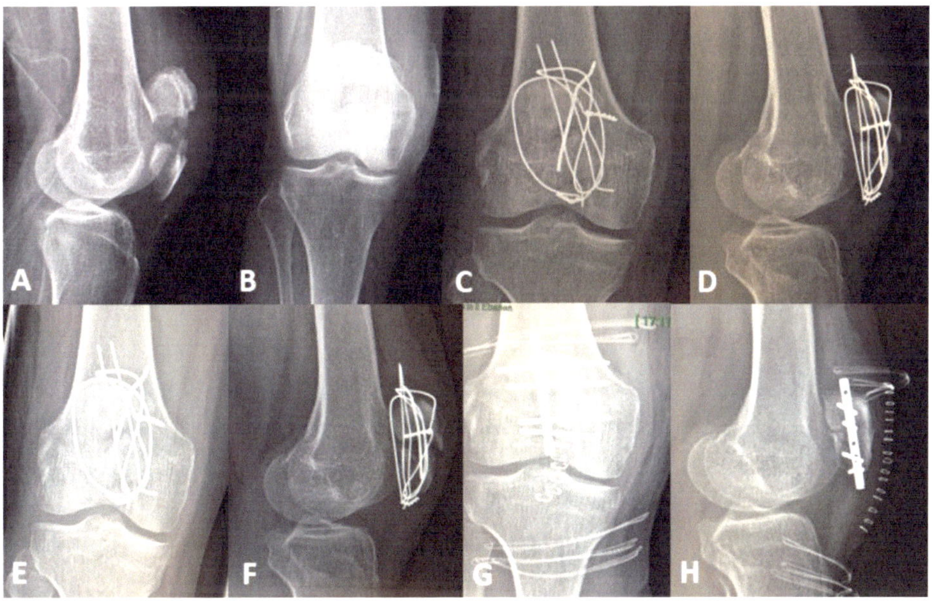

Figure 14.
(A and B) AO 34-C1 fracture of the patella; (C and D) internal fixation with tension band and cerclage wire; (E and F) failure of the construct tension band and cerclage wire; and (G and H) revision surgery with XS nail osteosynthesis.

As a result, an implant that allows compression of the entire fracture surface was needed. To achieve this, the XS nail entitled Tension Band Compression Nail (TCN) was developed and placed centrally in the patella; this implant allows equally distributed compression on the whole surface by muscular distraction [35] (**Figure 15**).

In a biomechanical study, the authors divided 30 sawbone patellas into four types of fractures. Proximal third, middle, distal third, and Y-pattern fractures were obtained by osteotomy. Osteosynthesis was carried out using one XS nail, two XS nails, and standard AO tension band. The three-part Y-pattern fractures were fixed with an additional circular wire system (cerclage) in the AO tension band group (**Figure 16**).

Plastic and total deformations were recorded, while the sawbone patellas were submitted to a force of 250 and 500 N at 30°. The batch with XS nail sets the lowest value in the entire lot. The highest deformation occurred in the AO tension band group, and a significant gap appeared between the fragments. No gap was registered in the XS-nail group.

Smaller differences were recorded in the Y-pattern group based on the strength of the circular wire put in close contact with the bone that was added to the tension band group. We must mention that in real clinical cases this is not possible because of the soft tissue interposition.

The experimental data show that the XS nail system is a viable alternative to the AO tension band due to its good fixation of the fragments and less deformation under physiological loading. This implant's characteristics will allow patients to recover faster and with better long-term results.

Tension band wiring of olecranon fractures has been the standard choice since 1963. However, migrations of the Kirschner wires and cerclage failure were reported in up to 80% of the cases [36]. Even a numerical model of the tension band wiring technique proved that high von Mises stresses were seen at the bridge between two fragments connected by the Kirschner wire [37].

The plate osteosynthesis alternative may aggravate a soft tissue lesion, usually caused by direct trauma.

The intramedullary XS nail is a new form of osteosynthesis that allows uniform compression of the fracture surfaces by central positioning the implant. Moreover, increased fixation in the cortical bone can be obtained by locking the 4.5 mm nail with threaded 2.4 mm wires. The surgical technique is easy and protective for the soft tissues due to the intramedullary position.

In a clinical study from 2006, 80 olecranon fractures were treated with XS nail osteosynthesis, and after a follow-up period of 16 months, the results were good in 93.2% of the cases [38] (**Figure 17**).

The XS nail represents, in our opinion, a future possibility in the treatment of olecranon fractures, especially in the elderly population, where bone quality is deficient.

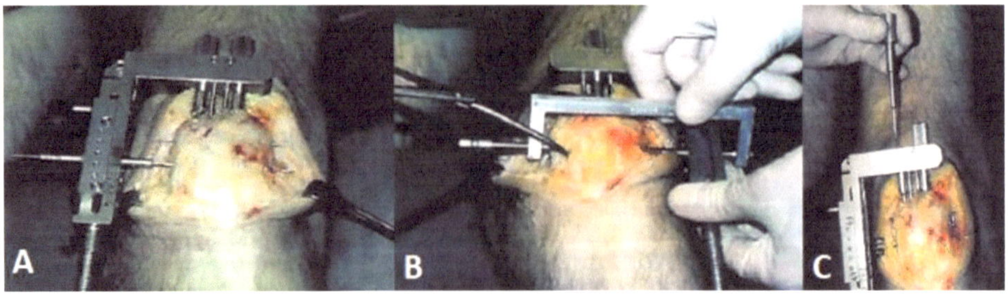

Figure 15.
(A) TCN with locking aiming device; (B) Kirschner wire aiming device; and (C) compression screw placement.

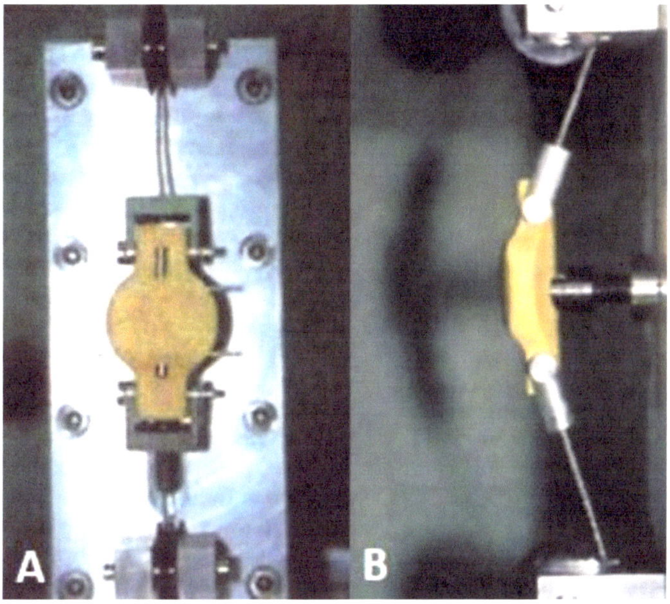

Figure 16.
Patella test without osteotomy in the hydraulic machine: (A) frontal view and (B) lateral view.

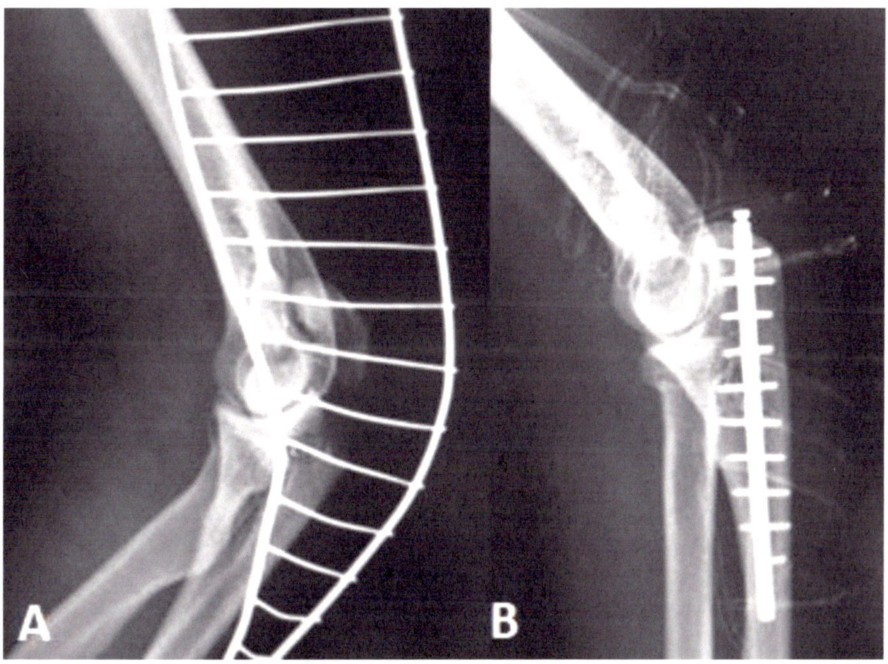

Figure 17.
(A) Displaced fracture of the olecranon and (B) osteosynthesis with XS nail.

7. Clinical and biomechanical studies regarding plates with polyaxial stability for fractures of distal radius and proximal humerus

Among the elderly, distal radius is the second most common fracture location after hip fractures. Mauck et al. [39] presented a wide variety of fracture patterns. Dorsal and metaphyseal radius fractures are usually treated with closed reduction and cast immobilization or K-wire fixation. However, the fracture is most often unstable, and the reduction is not always maintained. Therefore, all unstable fractures or articular fractures have to be treated surgically. Due to the stability deficiency of the osteosynthesis with standard plates, depending on the type of fracture, an additional ventral, dorsal, or even radial fixation with a high degree of complexity is necessary. Studies showed that this type of osteosynthesis was accompanied by complications related to tendons lesions and a high risk of secondary dislocation or angulation.

Since the 1990s, plates with angular stability started to be used more often in the treatment of distal radius fractures. Initially being used for simple metaphyseal fractures, angular stability plates were later used in complex fractures depending on the fracture pattern. Different types of screws had to be adapted to the different articular fragments, which needed to be fixed. Moreover, an articular positioning of the screws had to be avoided, and this was followed by unavoidable issues regarding stability [40].

Thus, multidirectional (polyaxial) plates with multiple angulation possibilities were developed. The screws can be inserted perpendicularly on the plate surface or at an angle of 10° distal/proximal, medial, or lateral, offering the possibility for insertion at plate level in positions very close to the articular surface. Depending on the bone structure and the multiple fracture trajectories, fixation of fracture fragments is always possible in the adequate position (**Figure 18**).

Because these implants require large incisions with soft tissue damage and deperiostation, a new type of implant was developed [41]. The XS radius (XSR) nail is a 4.5- or 3.5-mm straight nail that is introduced after drilling and inserting a guide wire inside the medullary canal. It is then locked using threaded wires in three different directions (**Figures 19** and **20**).

The authors tested the osteosynthesis with angular stable plate and XS nail on 16 osteotomized sawbones that replicate AO/A3 fractures of the distal radius. We registered the deformation after we subjected them to 1000 alternating load cycles from 20 to 200 N (**Figures 21** and **22**).

The experimental study showed a reduced deformation of the XS nail system compared to the plates with angular stability. The deformation amplitude was only 0.31 mm in the XS nail system compared to 0.42 mm in the angular stability plate [41].

Although both implants showed good biomechanical results, the deformation recorded in the XS group proved to be 20% lower than the plate group. The XS nail has the advantages of a simple operation technique, the intraosseous positioning, and saving the pronator quadrates; however, very comminuted fractures are better treated with multidirectional (angular) locking plate [40, 41].

To date, no consensus has been reached regarding the optimal treatment of proximal humerus fractures [42]. Instability and fragment displacement usually require surgical treatment for a better quality of life. Highly comminuted fractures

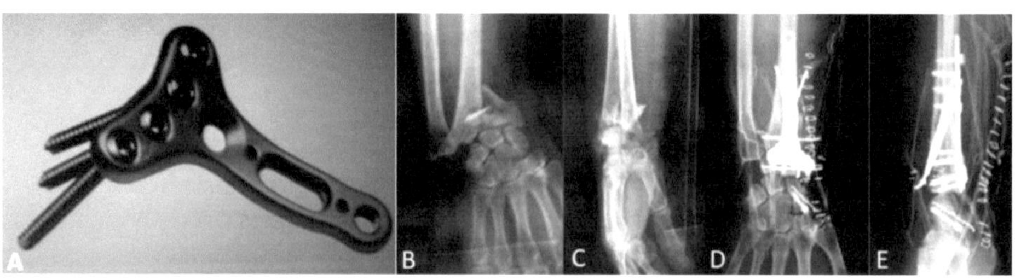

Figure 18.
(A) Polyaxial stability plate; (B and C) AO/C2 type distal radius fracture, carpal scaphoid fracture; (front and lateral view); (D and E) reduction and osteosynthesis of radius fracture with polyaxial stability plate + additional K wires and osteosynthesis of the carpal scaphoid fracture with Herbert screw + K wire.

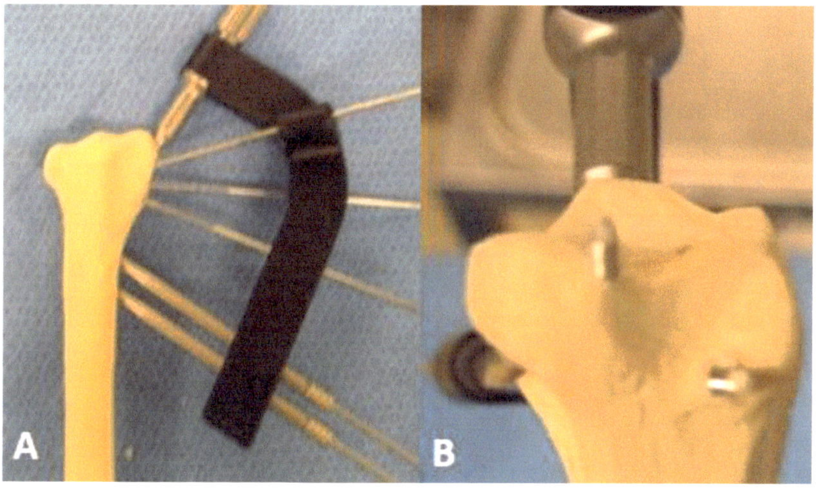

Figure 19.
(A) XS nail in sawbone with aiming device and (B) multidirectional locking of the XS nail with threaded K wires.

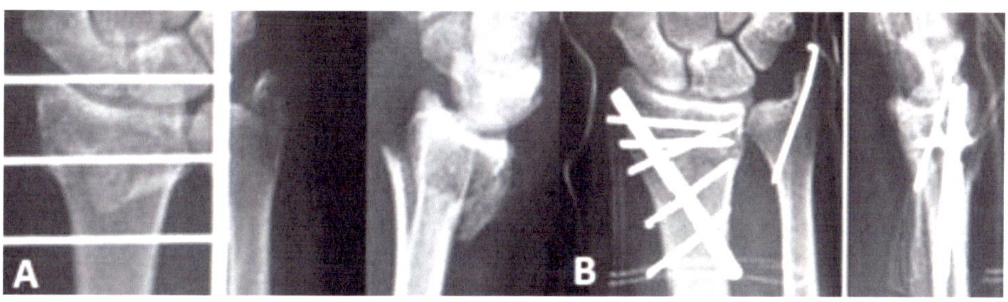

Figure 20.
(A) X-ray view of AO/A3 proximal radius fracture and (B) reduction and osteosynthesis with XS nail.

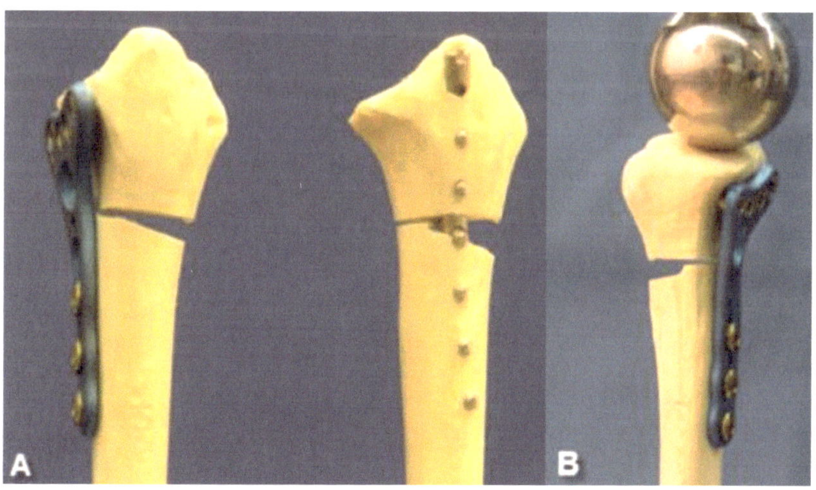

Figure 21.
(A) Angular stable plate and XS nail osteosynthesis and (B) alternating load test.

will necessitate hemi or total shoulder arthroplasty, but the vast majority of fractures can be treated by osteosynthesis with multidirectional angular stability plates or other types of plates (**Figures 23** and **24**).

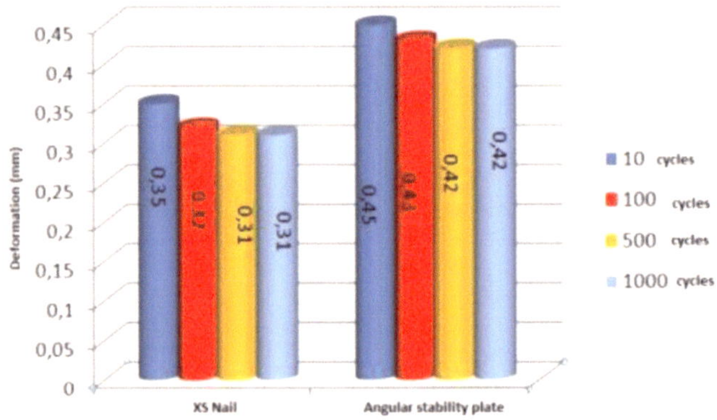

Figure 22.
Deformation values under load are higher for the plate than the XS nail.

Figure 23.
Locked polyaxial plate for proximal humerus.

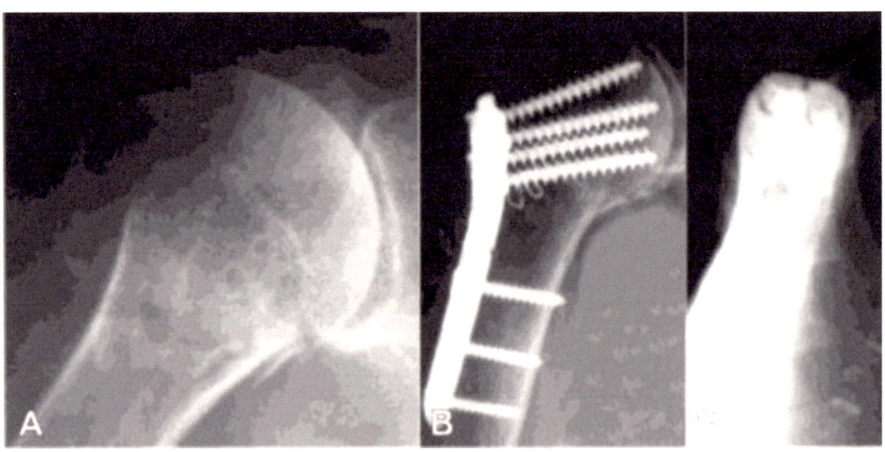

Figure 24.
(A) Proximal humerus fracture type AO 11-B1 and (B and C) locked polyaxial plate fixation.

The angular stable fixation of proximal humerus fractures has significantly improved the possibilities of anatomical reconstruction and postsurgical rehabilitation. Optimal screw positioning is hard to be achieved using classic T plates or monoaxial locked plates, especially in large bone defects or osteoporotic bones (**Figures 25** and **26**).

In a biomechanical study from 2006, the authors [43] compared on 18 sawbones the efficiency of osteosynthesis with multidirectional angular stability plate versus

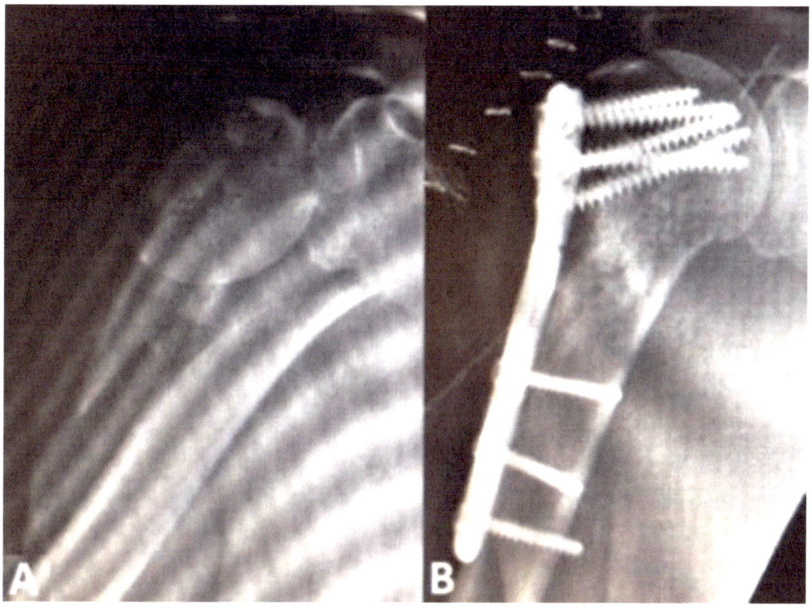

Figure 25.
(A) Proximal humerus fracture type AO 11-B3 and (B) fixation with polyaxial stability plate.

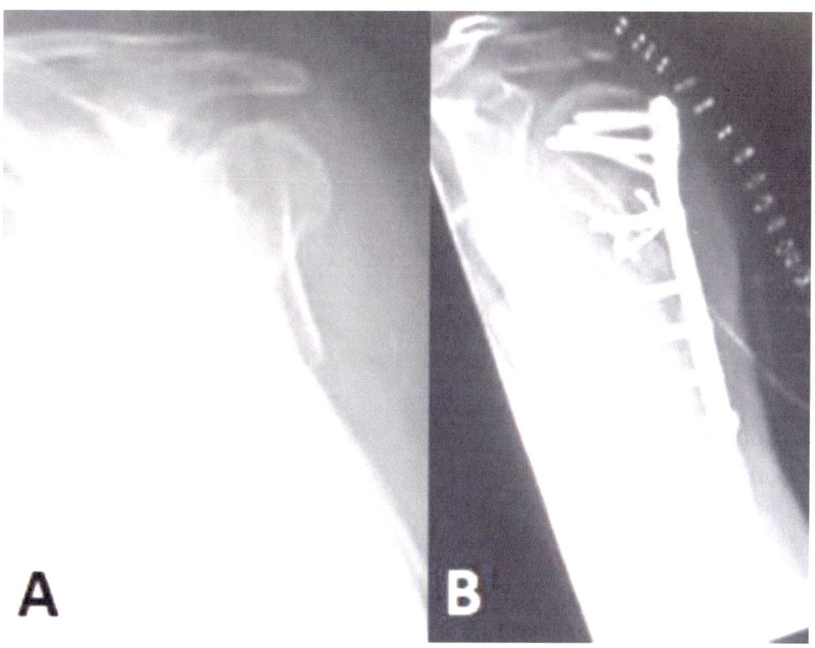

Figure 26.
(A) Proximal humerus fracture type AO 11-B3 fracture and (B) fixation with polyaxial stability plate.

monodirectional angular stability plate and normal (classic) T plate, with the same configuration and thickness. The bone-implant constructs were subjected to an alternating pressure load test of 1000 cycles between 50 and 200 N (**Figures** 27 and 28).

After 1000 cycles, the total deformation was 0.7 mm in the PAS group, 1 mm in the MAS group, and 1.5 mm in the T-plate group. In all the tests, the highest resistance and the lowest deformation were seen in the polyaxial stability plate group.

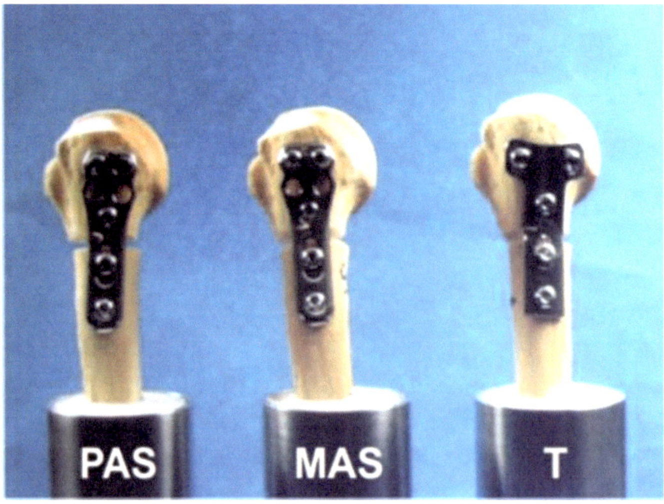

Figure 27.
Sawbones with polyaxial angular stability (PAS) plate, monoaxial plate (MAS), or classic T plate (T).

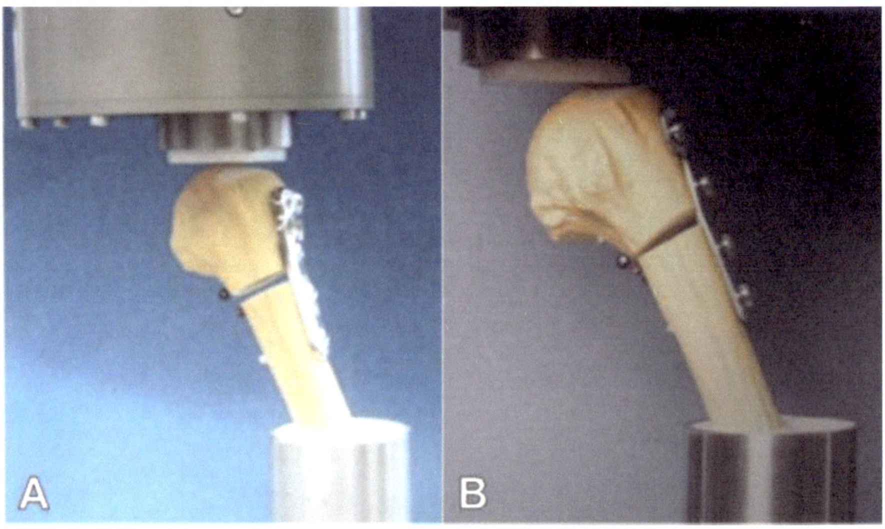

Figure 28.
(A) No deformations at the PAS plate under load and (B) the closing of the osteotomy segment at T plate under load.

In conclusion, the polyaxial angular stability plate offers not only an improved placement option for the screws but also a lower risk of reduction loss than the monoaxial plate or T plate [43].

8. Conclusions

The authors enhance the advantages of these innovative implants for difficult clinical trauma cases. For supracondylar fractures simulated on plastic composite bones, DCS is better than CBP in most loading tests. In complex midshaft humeral fracture, the shorter locked plate (LCP—Synthes) seems to be the most rigid implant; the intramedullary nail proved to be the most elastic, while the DCP

DOI: http://dx.doi.org/10.5772/intechopen.91728

gives surprising values of torsion forces relatively close to the longer locked plate (AxSOS—Stryker). Clinical and biomechanical studies revealed the superiority of intramedullary gliding nail over DHS and gamma nail due to the double-T blade profile. The XS nail is a secure device for ankle and pilon fractures; in osteoporotic bone and difficult soft tissue conditions, it shows significant advantage over the plate fixation. In fractures of the patella and olecranon, the XS nail allows uniform compression of the fracture surface and overcomes the disadvantages of the AO tension band. Clinical and biomechanical tests proved the superiority of the polyaxial locked plates in distal radius and proximal humerus fractures due to the adjustable trajectory of the screws; in distal radius, the XS nail is stronger than the plates.

Author details

Paul-Dan Sirbu[1]*, Wilhelm Friedl[2], Dan Mihailescu[1], Liliana Savin[1], Andrei Scripcaru[1], Norin Forna[1], Mihnea Theodor Sirbu[1], Mihaela Pertea[1] and Razvan Cosmin Tudor[3]

1 Grigore T. Popa University of Medicine and Pharmacy Iasi, Romania

2 Rotkreuzklinik Wertheim, Germany

3 Vaslui County Hospital, Romania

*Address all correspondence to: pdsirbu@yahoo.com

References

[1] Koval KJ, Hoehl JJ, Kummer FJ, Simon JA. Distal femoral fixation: A biomechanical comparison of the standard condylar buttress plate, a locked buttress plate, and the 95-degree blade plate. Journal of Orthopaedic Trauma. 1997;**11**(7):521-524. DOI: 10.1097/00005131-199710000-00010

[2] Marty A, Fankhauser C, Frenk A, Cordey J, Gasser B. Biomechanical evaluation of the less invasive stabilization system for the internal fixation of distal femur fractures. Journal of Orthopaedic Trauma. 2001;**15**(7):482-487. DOI: 10.1097/00005131-200109000-00004

[3] Higgins TF, Pittman G, Hines J, Bachus K. Biomechanical analysis of distal femur fracture fixation: Fixed-angle screw-plate construct versus condylar blade plate. Journal of Orthopaedic Trauma. 2007;**21**:43-46. DOI: 10.1097/BOT.0b013e31802bb372

[4] Heiney J, Barnett M, Vrabec G, Schoenfeld A, Baji A, Njus G. Distal femoral fixation: A biomechanical comparison of trigen retrograde intramedullary (IM) nail, dynamic condylar screw (DCS), and locking compression plate (LCP) condylar plate. The Journal of Trauma. 2009;**66**:443-449. DOI: 10.1097/TA.0b013e31815edeb8

[5] Jaakkola JI, Lundy DW, Moore T, Jones B, Ganey TM, Hutton WC. Supracondylar femur fixation. Mechanical comparaison of the 95° condylar side plate and screw versus 95° angled blade plate. Acta Orthopaedica Scandinavica. 2002;**73**(1):72-76. DOI: 10.1080/000164702317281440

[6] Sirbu PD, Carata E, Petreus T, Munteanu F, Popescu C, Asaftei R, et al. Dynamic condylar screw (DCS) versus condylar blade plate (CBP) in complex supracondylar femoral fractures–a biomechanical study. IFMBE Proceedings, International Conference on Advancements of Medicine and Health Care through Technology. 2009;**26**:409-412. DOI: 10.1007/978-3-642-04292-8_90

[7] Jeffrey K, Booth G, Day R. A biomechanical comparison of locking plate and locking nail implants used for fractures of the proximal humerus. Journal of Bone and Joint Surgery. 2007;**16**:362-366. DOI: 10.1016/j.jse.2006.01.019

[8] Sanders BS, Bullington AB, McGilivaryGR,HuttonWC.Biomechanical evaluation of locked plating in proximal humeral fractures. Journal of Bone and Joint Surgery. 2007;**16**:229-234. DOI: 10.1016/j.jse.2006.03.013

[9] Dunlap JT, Chong AC, Lucas GL, Cooke FW. Structural properties of a novel design of composite analogue humeri models. Annals of Biomedical Engineering. 2008;**36**(11):1922-1926. DOI: 10.1007/s10439-008-9568-y

[10] Ulian V, Mazzer N, Barbieri C, Moro CA, Oliveira LA. Primary stabilisation of humeral shaft fractures: an experimental study of different osteosynthesis methods. Acta Ortopédica Brasileira. 2008;**16**:8-12. DOI: 10.1590/S1413-78522008000100001

[11] Asaftei R,Sîrbu PD, Carata E, Bar M, Botez P. Biomechanical analysis of three different types of implants in humeral diaphysis fractures. In: Proceedings Atequal: Advanced Technologies for Enhanced Quality of Life; IEEE 2010, 14-17; 2010. ISBN: 978-1-4244-8842-1. DOI: 10.1109/ATEQUAL.2010.35

[12] Kaufer H. Mechanics of the treatment of hip injuries. Clinical Orthopaedics and Related Research. 1980;**146**:53-61

[13] Kyle RF, Gustilo RB, Premer RF. Analysis of six hundred and twenty-two intertrochanteric hip fractures. The Journal of Bone and Joint Surgery. American Volume. 1979;**61**:216-221

[14] Baumgaertner M, Custin S, Lindskog D, Keggi J. The value of the tip-apex distance in predicting failure of fixation of peritrochanteric fractures of the hip. Journal of Bone and Joint Surgery. 1995;77:1058-1064

[15] Baumgaertner MR, Solberg BD. Awareness of the tip-apex distance reduces failure of fixation of trochanteric fractures of the hip. Journal of Bone and Joint Surgery. British Volume (London). 1997;**79**:969-971

[16] Suckel A, Helwig P, Schirmer A, et al. Komplikationsraten bei der Versorgung von der per-und subtrochanteren Femurfrakturen mit zwei intramedullaren Osteosynthesenverfahren. Vergleich eines konventionellen Nagelsystems mit einem rotationsstabilen Verfahren im Kopf-Hals-Fragment, Gammanagel und Gleitnagel. Zentralbl Chir. 2003;**128**(3):212-217. DOI: 10.1055/s-2003-38535

[17] Gehr J, Arnold T, Hilsenbeck F, et al. The gliding nail, a universal implant in the treatment of proximal femur fractures. European Journal of Trauma. 2006;**32**:562-569. DOI: 10.1007/s00068-006-5117-2

[18] Friedl W. The gliding nail. A new implant for complication-free primary load-bearing management of per and subtrochanteric femoral fractures. Langenbecks Archiv für Chirurgie. Supplement. Kongressband. 1996;**113**:970-973

[19] Friedl W, Anthoni C, Fritz T, Schmotzer H, Wipf M. Importance of blade geometry for stability of fixation with short intramedullary nailing systems for the proximal end of the femur (gliding nail). Langenbecks Archiv für Chirurgie. Supplement. Kongressband. 1998;**115**:1224-1226

[20] Friedl W, Göhring U, Fritz T, et al. Die Gleitnagelosteosynthese Ein neues universell einsetzbares Implantat zur Versorgung per- und subtrochanterer Femurfrakturen. Der Chirurg. 1998;**69**:191-197. DOI: 10.1007/s001040050395

[21] Friedl W, Hilsenbeck F, Sturzenhofecker P. The gliding nail (GN). An universal implant for immediate weight bearing restauration in all per- and subtrochanteric femur fractures. Evaluation of 501 patients. Advanced Technologies for Enhanced Quality of Life. 2009:61. DOI: 10.1109/AT-EQUAL.2009.41

[22] Phillips WA, Schwartz HS, Keller CS. A prospective, randomized study of the management of severe ankle fractures. Journal of Bone and Joint Surgery. AmericanVolume. 1985;**67**:67-78. DOI: 10.2106/00004623-198567010-00010

[23] Collinge CA, Heier K. Ankle fractures and dislocations. In: Stannard JP, Schmidt AH, Kregor PJ, editors. Surgical Treatment of Orthopaedic Trauma. Germany: Thieme; 2007. pp. 792-814

[24] Batten TJ, Scott-Davies C, Butler M, Parsons SW, Walter RP. Percutaneous screw fixation for painful non-union of lateral malleolus ankle fractures. Injury. 2018;**49**(10):1936-1941. DOI: 10.1016/j.injury.2018.08.010

[25] Chiang CC, Tzeng YH, Lin CC, Huang CK, Chang MC. Minimally invasive versus open distal fibular plating for AO/OTA 44-B ankle fractures. Foot & Ankle International. 2016;**37**(6):611-619. DOI: 10.1177/1071100715625292

[26] Zaghloul A, Haddad B, Bakersfield R, Davis B. Early complications of surgeryin operative

treatement of ankle fractures in those over 60: A rewiew of 186 cases. Injury. 2014;**45**:780-783. DOI: 10.1016/j.injury.2013.11.008

[27] Gehr J, Neber W, Hilsenbeck F, Friedl W. New concepts in the treatment of ankle joint fractures: The IP-XS (XSL) and IP-XXS (XXSL) nail in the treatment of ankle joint fractures. Archives of Orthopaedic and Trauma Surgery. 2004;**124**:96-103. DOI: 10.1007/s00402-003-0606-9

[28] Sirbu PD, Tudor R, Scripcaru A, Veringa V, Sirbu MT, Forna N, Pertea M. Locked intramedullary XS nail for ankle fractures. Design, biomechanics and preliminary results. In: E-Health and Bioengineering Conference (EHB) 2019; Iasi, Romania; 2019. pp. 1-4. DOI : 10.1109/EHB47216.2019.8969958

[29] Mb O, Aksan T, Ertekin C, Tezcan M. Coverage of exposed bone and hardware of the medial malleolus with tibialis posterior artery perforator flap after ankle fracture surgery complications. International Wound Journal. 2019:429-435. DOI: 10.1111/iwj.13289

[30] Gehr J, Friedl W. Intramedullary locked fixation and compression Nail (XS nail). Treatment of ankle fractures. Open OrthopTraumatol. 2006:155-170. DOI: 10.1007/s00064-006-1168-0

[31] Gehr J, Friedl W. New concept in the treatment of distal metaphyseal and little dislocated pilon fractures with soft tissue damage. Unfallchirurg. 2002;**2**:520-532. DOI: 10.1007/s00113-001-0395-y

[32] Gehr J, Hilsenbeck F, Arnold T, Friedl W. Minimally invasive management of distal metaphyseal tibial fractures and pilon fractures. European Journal of Trauma. 2004;(6):378-386. DOI: 10.1007/s00068-004-1414-9

[33] Friedl W, Gehr J. Intramedullary/intraosseus osteosynthesis of pilon and ankle fractures. In: Advanced Technologies for Enhancing Quality of Life, Iasi; 2010. pp. 3-5. DOI: 10.1109/ATEQUAL.2010.25

[34] Klute I, Meenen NM. Die Fraktur der Kniescheibe. Heidelberg. New York: Springer; 1998

[35] Friedl W, Clausen J. Experimental examination of a tension-compressio n-nail (XS) in the osteosynthesis of patella fractures. Osteosynthese International. 2000;**8**:252-260

[36] Finsen V et al. AO tension band osteosynthesis of displaced olecranon fractures. Orthopedics. 2000;**23**:1069-1071

[37] Greenfield JRF et al. A numerical model of the tension band wiring technique for olecranon fracture reduction. Applied Mathematics and Computation. 2016;**297**:31-38. DOI: 10.1016/j.amc.2016.10.016

[38] Gehr J, Friedl W. Intramedullary locking compression nail for the treatment of an olecranon fracture. Operative Orthopädie und Traumatologie. 2006;**3**:199-213

[39] Mauck BM, Swigler CW. Evidence-based review of distal radius fractures. Orthopedic Clinics of North America. 2018;**49**(2):211-222. DOI: 10.1016/j.ocl.2017.12.001

[40] Sirbu PD, Friedl W, Botez P, Stratan L, Hopulele S, Asaftei R. Minimally Invasive Plate Ostehsynthesis – Internal Fixators. Iasi: Casa de editura Venus; 2008. ISBN: 987-973-756-086-5

[41] Friedl W, Rinner M, Simnacher M, Mathieu C. Angle Stable plate or angle stable locked nail fixation in distal radius fractures: An experimental examination and phase I clinical study. In: Advanced Technologies for Enhancing Quality of Life; 2010. ISBN:

978-1-4244-8842-1. DOI: 10.1109/ATEQUAL.2010.26

[42] Klug A, Wincheringer D, Harth J, Schmidt-Horlohé K, Hoffmann R, Gramlich Y. Complications after surgical treatment of proximal humerus fractures in the elderly - an analysis of complication patterns and risk factors for reverse shoulder arthroplasty and angular-stable plating. Journal of Shoulder and Elbow Surgery. 2019 Sep;**28**(9):1674-1684. DOI: 10.1016/j.jse.2019.02.017

[43] Friedl W, Nan H. Mono-versus Multidirektional Winkelstabile Plattenosteosynthese bei proximalen Humerusynthese. Eine Experimentelle Untersuchung. Deutscher Kongress für Orthopädie und Unfallchirurgie. 70. Jahrestagung der Deutschen Gesellschaft für Unfallchirurgie, 92. Tagung der Deutschen Gesellschaft für Orthopädie und Orthopädische Chirurgie und 47. Tagung des Berufsverbandes der Fachärzte für Orthopädie. Berlin, December 2-6, 2006. Düsseldorf, Köln: German Medical Science; 2006. DocW.4.8.6-461

Chapter 8

Kinematical Analysis of the Volleyball Auction in Preyouth and Youth Players of the Pichincha Sports Concentration

Erik David García Lasluisa
and Jefferson Michael Vela Rodríguez

Abstract

The finish is an essential element since it is a major action within offensive elements that help the team get points, so it is necessary to analyze each of the phases of this foundation. As such need is to analyze the different angles of movement, speed and center of mass to establish kinematic reference values of the gain shape values that can serve as a reference when the technique of the players so potentiate the driving habit means is analyzed optimizing sports training at all levels. The authors of this study aimed to biomechanically analyze the phases of the auction volleyball in prejuvenile and young players of Pichincha sports concentration. They studied 19 volleyball players from the Pichincha sports concentration under non-probability purposive sampling, with 11 athletes from the prejuvenile team and 8 youth team athletes. Kinovea to measure software (experimental version) was used and the U test was used Mann-Whitney SPSS d e13 digital camera megapixels, measuring tape and HP computer with Intel Core i5 processor. In the approach phase, there was no significant difference in regard to the column angle ($p = 0.457$). In whipped phase, no significant difference in column angle ($p = 0.283$). In the preparation phase for beating on the angle of the elbow, there is no significant difference ($p = 0.83$). In the striking phase at the angle of the elbow if there is significant difference ($p = 0.002$). In the decay phase in the column angle no significant difference ($p = 0.83$). This was established with the Mann-Whitney U test. Column angles do not significantly influence the auction technique. However, the influence of angles of the elbow was observed, for better elbow flexion and hand placement can generate a greater impact on the ball and therefore greater difficulty in opposing defense.

Keywords: biomechanics, auction, movement angles, center of gravity, volleyball

1. Introduction

The remante is a technical gesture of great importance in volleyball and is also the most important offensive element (attack) consisting of two previous actions before sending the ball: these are terminal and continuity within terminals we serve, block and remante [1]. This technique is one of the most decisive factors in athletic performance. In volleyball, it can produce more than 100 shots during a game by one player; so, the difficulty lies in combining skills with previous career jump and hit. Therefore, the auction should have a refined technique that allows executing proper moves to achieve economy and rationality of the effort [2]. One of the factors that stand out performance in volleyball is the force as it contributes to the physical appearance and efficiency of the game [3]. The effectiveness of the auction to determine success of the attack and the possibility of achieving several points suggests authors that 60% of the points are obtained through the auction where the core has a 78.8% efficiency pins 83.3%, opposite 75.7% and finally the offensive 62.9% effective in total offensive actions [4]. Within a party hop count it is 1802 of this amount 35% are breaks the remante [5].

The effectiveness of the technique in biomechanical terms is given to achieve the greatest possible height of the center of gravity in the boost phase, and hit the ball at the highest point maximum printing speed and accuracy at the time of the beating [6].

The auction requires explosive strength, speed, and accuracy [7]. This technical element involves special features because it is done without a base of support such that parts of the lower body are completely suspended in the air to hit the ball; therefore, the center of gravity tends to change and it should be emphasized that 100% of the time it takes the shot exists opposition adversary (lock) [8] therefore jump over the network and project the ball as much force does not guarantee you get points because factors such as skill, coordination and precision hand oculus also complement the movement. From the point of view of biomechanical studies, analysis of the auction aims to provide relevant information about the kinematic and kinetic results of the implementation of the technical element (auction) and thus to establish movement patterns whose aspects are expressed to be corrected during the learning stages of the auction and the correlation they have with the ideal models, that is more experienced players to correct finally identifies movements that can be the cause of injury during sports [9].

Moreover, the auction consists of two parts: the jump and hit, where the jump aims to achieve the maximum height of the center of gravity because it allows exceed block the opponent and finally the goal of beating is to get as much strength, speed, and accuracy in order to reduce the response time of the opponent [10]. In the phase of the jump one can see that the body tends to bend forward at the waist [11]. Applying biomechanical laws to this movement, we can say that he achievement not hit the ball with maximum force because the existence hip flexion biokinetic chain does not activate all levers, and according to Newton's second law the acceleration acquired by a body is proportional to the force applied to it; that is, in the auction the player, he must stay with his motionless body during impact [11].

The auction is divided into five phases: 1st approach stroke, Whipped 2nd, 3rd Preparation of beating, 4th and 5th Fall beating. The first two phases aim to achieve the maximum height of the center of gravity through the jump in the following two phases, which seek to achieve the right ball speed, strength, and direction; finally, the last phase seeks to cushion the body segments after the jump [10] (**Figure 1**).

During the advance stroke the player reaches between 50 and 60% of its maximum speed [13] since its goal is to produce the greatest amount of movement to the beating or biomechanical terms generate the horizontal movement needed to transform it into vertical movement (jump) plus a career boost generates effective 20% more height on the jump, this is the performs diagonally to the grid because in zone IV oblique race 45 is described to 60° while in zone II the path is straight with an angle of 60–90° because the player must finish off the network [8]. Travel speed and the number of steps required to reach the ball will depend on anthropometric factors of the player such as the stride length and angles of movement. At this stage, the final step becomes the most important and, therefore, should be broader and faster, allowing slow down of the other foot to generate a parallel supporting foot (heel, sole, and toe), culminating in the knee flexion; the time taken for this is approximately 0.17–0.19 and most commonly used muscles in motion are: quadriceps, buttocks, adductor, biceps femoris, twins, hamstrings, and erectors of the spine (**Figure 2**).

Phase whipped aims leave the player with the shoulder of the arm that performs the beating ahead of the ball; it lasts for an approximate time of 0.13–0.18 and the most commonly used muscles are: trapezius, deltoids, biceps, pectoral in the upper body and quadriceps, glutes, adductors, hamstring, calves, hamstrings and erector spine in the lower body [8]. It must be taken into consideration that during this phase, the body weight is transferred to the heels, then to the soles of the feet, and ends at the tips that accompany rapid and strong arm movements, ending in the explosive spread of the train generating lower takeoff [14]. During beaten two periods where there support and a subsequent double support because these let you apply as much force to raise the center of gravity, require the coordination of arms to produce more vertical speed and increase power occurs the jump which will allow to have a height greater flight and elevation to the time of the beating [1]. This technical element depends on the skill of the player as well as physical aspects such as the explosive force of the lower body, neuromuscular coordination, and genetic traits like height; however, the generate an optimum amount of vertical speed (feed stroke) height is maximum.

Figure 1.
Phases sequence shot [12].

Figure 2.
Approach race.

In boosting whipped extension joints of the concentric and eccentric generate trunk contractions powertrain causing the decline of the center of gravity and then release energy and raise the center of gravity in addition to the accompany him to the swing arm both the boost phase and prior to increases braking 54% vertical speed during the jump should be noted that the jump support is more common in women with a 16% incidence in the game although according to several authors twice support generates more height but requires more time [13]. "The angle of the (right) foot forward corresponds on average to 137 degrees while the left leg angle of 162 degrees. That is, while the member that one time one-two is placed in a moderate bending, two time member is placed in greater flexion" [15] (**Figure 3**).

Preparation for beating is the phase in which the feet lose contact with the ground and the shoulder hitting the ball reaches its maximum height [16]. In this phase, the players are airborne and must perform movements that allow generation of the maximum speed of the ball striking the hand, where three phases are involved: preparation, assembly, and acceleration. During preparation, the feet leave the ground and shoulder internal rotation occurs; in the stage of assembly of the shoulder, external rotation starts and ends at the start of the internal rotation as the arm strikes; finally the acceleration is the phase in which the internal rotation of the shoulder ends with the contact of the ball, which will have the speed with which it was impacted and this is around 69.2 km/h—it should be emphasized that this is not the maximum speed of the auction but the speed with which the ball arrives at impact [17] the hand does not hit the ball follows the trajectory of the arm forward and up, at the same time the elbow of the arm that performs the beating describes the same path hand accompanied by the elevation of the head and bending the movements legs causing a bow. As for the speed with which the hand hits the ball, it depends 46% extension in 20.5% of the rotation of the shoulder, 14.5% of the performance of the trunk rotation, by 7, 5% forward displacement of the center of gravity during the jump, 5.5% of wrist flexion and 6% remaining dependent on other factors [8] (**Figure 4**).

Figure 3.
Beat.

Figure 4.
Preparation for the hitting.

Phase of scrimmage begins at the moment when the hand that takes the blow comes into contact with the ball and ends when losing contact with it; usually an angle of 140–170° is formed in the shoulder. A the ends movements acceleration phase product elbow extension which ends the movement of the arm describes the path following occurs: forward and downward, further movement compensation of the where the most important undercarriage are performed: extension knee and rotation of the trunk [8]. In this phase, the shoulder should reach the maximum height until the hand contacting with the ball is at its highest point [16]; the steps involved in the kinetic chain are: hip rotation around the vertical axis, displacement and rotation of the trunk, flexion and rotation of the shoulder, elbow extension and pronation of the forearm and wrist flexion [18].

From the point of view of biomechanics, the remante to the reach the stage meets scrimmage with two goals: the first is to reach the maximum height, which is a function of the anatomical and physical capabilities of each player. The second is to reach full speed and this will depend on the position of the hand when it hits the ball. So, having separate fingers (cup position) ensures a path with maximum speed and right direction; also, the body must be fully extended and perpendicular to the floor being located slightly ahead of the shoulder arm hit factors dependent approach stroke and the place where the clay is performed (jump) [14]. Muscles that most influence the movement are: trapezius, deltoid, pectoral, subscapular, prickly infra, supraspinatus, biceps, triceps, flycatchers, abdominal, and teres major quadriceps (**Figure 5**).

The phase of the drop starts when the hand hits the ball descend in front of the body and due to gravity the player descends into contact with the ground, which takes a cushioning movement through knee flexion [19]. The support for body is through both legs (damping) involving joint ankle, knee, and hip in the order of the parts of the body in contact with the ground to end the balance fall through pendulous movements made by the opposite arm, which carries the blow [18]. The purpose of this phase is to reduce the stress of the repetitive impact of the body against

Figure 5.
Hit.

the ground to protect the joints of the ankle, knee, hip, and spine from injury. Also, the fall must be performed with two supports (2 ft) to divide the force of the impact (9.8 m/s by body weight); the muscles involved in the movement are: soleus, twins, hamstrings, erector spinae, rectus abdominis, dorsal, deltoids, biceps, and triceps [8] (**Figure 6**).

Biomechanics is the study of the application of laws and mechanical foundations of biology, especially structures of the musculoskeletal system and muscle activity; it also examines the internal and external forces acting on the human body and the effects they produce [20]. To make qualitative or quantitative analysis, one should consider three aspects: movement control, the structure of the body moving, and strength are already external and internal [21]. Finally biomechanics can establish the principles of training patterns of movement, to achieve the optimal technique taking as a starting point variability bilógicos systems (changes in athletes as the age maturation of the neuronal system, bone growth, evolution physical capabilities) and thus establish ranges that determine the level and stability domain of a technical model [22]. Furthermore, kinematics is the spatially detailed study of the movement of bodies, based on the displacements, whose velocities, accelerations, and forces cause the movements that are present in said displacements [23]. The locomotor allows kinematic analysis of the body through capturing movement then deducing equations. In addition, the musculoskeletal system is analyzed and as delivery this force to accelerate and decelerate the limbs to generate movement [24]. The bio-cinemática chain is successively joining a number of pairs biocinemáticos, whether open or closed [25]. From the point of view of engineering, it is a closed system in which every movement has a relationship; therefore, it ensures that all forces are transmitted positively and by default. Meanwhile, in a living organism, it is an open system where its dimension is determined by the linear distance from a hinge axis to another; it comprises muscle mass, bone structure, and the type of joint [26]. A biokinetic pair involves two bone members moving together where the chances of the movement depend on the structure of the union and the influence of muscles [25], while an open kinetic chain comprises of movements where the distal segment (hands or feet) moves freely in space with the body staying at a fixed point [27].

Figure 6.
Fall.

While the angle of movement is a way of providing a position; while the angle change is a distance covered by an angular movement from the point of view of Fisca activity is considered to angular movements made by a segment of the body relative to the hinge that is performed [28]. Finally, the center of gravity of a rigid body is the point of equilibrium, so that all the forces acting on the body are balanced, that is, their sum is equal to 0. For a flexible body, the body's center of gravity changes depending on the form of the body. For a person who is standing, it is located right at the level of the second sacral vertebra on a vertical line that touches the ground about 3 centimeters ahead of the ankle joint. It is all-important to possess the ability to vary the position of the center of gravity to maintain balance while walking or athletic actions are executed [29].

Participants in the study carried the art capping front biomechanically analyzing the different phases of the technical gesture (**Tables 1–4**) using the angle of the elbow to the preparation phase of the beating and phase hit to the ball, also of the angle of the column for the remaining phases of the technical movement. The software Kinovea 8.15 (experimental release) was used for biomechanical measurements. Data collection was performed at the Coliseum volleyball Concentration Deportiva de Pichincha under situations reestablished with periods of complete rest (6 min between kick) and a total of three shots for each player. Data collection involved athletes performing three shots on the opposite field region 4, of which the best shot for the respective biomechanical analysis was chosen, using a Sony professional, pointing out landmarks of interest. The videos were recorded in 2 days: on the first day, videos of 11 athletes of the prejuvenile team were recorded and on the second day, those of 8 athletes from the youth team were recorded.

The comparison of the technical stages of the closing or attack volleyball were performed using SPSS v.22, using the Mann-Whitney U test ($p \leq 0.05$) for two independent samples to the absence of normal distribution of the data [30].

2. Methods

Probabilistic not intentional sampling was used, where we studied the population of athletes from the volleyball team of the Sports Concentration of Pichincha,

No.	Edad (años)	Medida (cm)	1st phase angulo columna	2nd phase angulo columna	3rd phase angulo codo	4th phase angulo codo	5th phase angulo columna	TTR	TPG	ACCG
1	16	28	161.50	154.90	33.00	215.00	176.20	2.05	0.20	170.8
2	15	25	156.40	151.80	80.00	249.00	172.80	1.95	0.17	178.3
3	16	27	165.30	162.00	85.00	215.00	171.60	1.85	0.27	172.4
4	16	26	151.90	162.60	335.00	209.00	141.20	1.98	0.13	178.4
5	15	24	144.90	157.80	90.00	211.00	176.80	1.74	0.13	173.3
6	15	25	162.50	168.20	68.00	185.00	171.10	1.71	0.20	178.9
7	16	26	140.20	160.60	45.00	207.00	159.80	2.28	0.24	175.6
8	15	26	154.50	172.20	68.00	192.00	165.00	2.25	0.20	170.3
9	16	26	143.80	175.30	75.00	218.00	179.80	1.95	0.20	163.2
10	15	25	148.50	169.30	73.00	220.00	173.40	1.85	0.24	177.2
11	16	24	161.00	176.30	29.00	221.00	163.40	1.78	0.24	169.8
Promedio			153.68	164.64	89.18	212.91	168.28	1.94	0.20	173.47

TTR: Tiempo Total del Remate; TPG: Tiempo de la Preparación hasta el Golpe; ACCG: Ángulo del cuerpo en relación al centro de gravedad.

Table 1.
Variables en deportistas prejuveniles.

No.	Edad (años)	Medida (cm)	1ra fase angulo columna	2da fase angulo columna	3ra fase angulo codo	4ta fase angulo codo	5ta fase angulo columna	TTR	TPG	ACCG
1	23	25	152.10	174.80	83.00	164.00	171.80	2.01	0.24	175.4
2	25	23	154.80	168.00	91.00	191.00	154.20	1.68	0.20	177.8
3	17	24	160.20	162.80	69.00	179.00	156.30	1.28	0.17	173.6
4	17	23	151.60	178.50	95.00	157.00	142.40	1.85	0.24	169.8
5	18	25	159.60	166.30	110.00	200.00	174.30	1.58	0.24	175.8
6	17	27	158.40	161.10	65.00	188.00	149.40	1.44	0.17	177.4
7	20	28	163.10	166.60	93.00	190.00	144.10	1.95	0.24	169.5
8	22	25	160.50	173.20	89.00	197.00	169.40	1.41	0.20	179.4
Promedio	157.54	168.91	86.88	183.25	157.74	1.65	0.21	174.84		

TTR: Tiempo Total del Remate; TPG: Tiempo de la Preparación hasta el Golpe; ACCG: Ángulo del cuerpo en relación al centro de gravedad.

Table 2.
Variables en las deportistas juveniles.

	Group	N	Average range	Sum of ranks
Angle column	Pre youth	11	13.36	147.00
	Youth	8	5.38	43.00
	Total	19		
Contrast statistics[a]	**Angle elbow**			
U de Mann-Whitney	7000			
W de Wilcoxon	43,000			
Z	−3057			
Sig. asintót. (bilateral)	.002			
Sig. exacta [2 × (Sig. unilateral)]	.001[b]			

[a]*Group variable: Group.*
[b]*Not corrected for ties.*

Table 3.
Elbow angle in the striking phase.

	Group	N	Average range	Sum of ranks
Auction time	Pre youth	11	12.36	136.00
	Youth	8	6.75	54.00
	Total	19		
Contrast statistics[a]	**Auction time**			
U de Mann-Whitney	18,000			
W de Wilcoxon	54,000			
Z	−2154			
Sig. asintót. (bilateral)	.031			
Sig. exacta [2 × (Sig. unilateral)]	.033[b]			

[a]*Group variable: Group.*
[b]*Not corrected for ties.*

Table 4.
Time it takes to complete the technical basis of the auction.

with 11 athletes from the prejuvenile team (16–17 years) and 8 athletes from the youth team (18–19 years), all females.

3. Results

Table 1 shows the data obtained from the 11 prejuvenile athletes where average 153.68° was obtained in the first phase; an average of 164.64° was obtained in the second phase; 89.18° in the third stage; an average of 212.91° in the fourth stage; and an average of 168.28° in the fifth phase. In the total execution time of the auction technique (TTR) averaged 1.94 s, while the preparation time until the stroke of the ball (TPG) was averaged from 0.20 s, yielding an average of 173.47° in the angle of the body in relation to the center of gravity at the time of impact with the ball.

Table 2 shows the data obtained from 8 juvenile athletes studied. We obtained an average of 157.54° in the first phase; 168.91° in the second phase; 86.88° in the third stage; 183.25° in the fourth stage; and an average of 157.74° in the fifth phase of the technical movement studied. In the total execution time of the auction technique (TTR) averaged 1.65 s, while the preparation time until the stroke of the ball (TPG) was averaged from 0.21 s, yielding an average of 174.84° in the angle of the body in relation to the center of gravity at the time of impact with the ball. **Tables 3** and **4** the results of statistical contrast values also shown that generated significant differences Mann-Whitney U ($p \leq 0.05$) samples.

Table 3 shows that the average range of the preyouth team is 13.36 and the average range of the youth team is 5.38; after analyzing in the SPSS program, it is observed that there is a significant difference since the value obtained is 0.002, which is less than 0.05.

Table 4 shows that the average range of the preyouth team is 12.36 and the average range of the youth team is 6.75. After analyzing in the SPSS program, it is observed that there is a significant difference since the value obtained is 0.031, which is less than 0.05.

4. Discussion

In the phase of stroke approximation n lower average range is evident in the áng ulo column group pre juvenile (9.18) obtained in the young group (11, 13) with no significant differences ($p = 0.492$). In the whisk the angle of the column in relation to the floor was evidenced with a lower average range in the preyouth group (8.82) than that obtained in the youth group (11.63), without significant differences being evidenced ($p = 0.310$), while the preparation phase for the beating of the same form showed a lower average range in preyouth (8.09) than in youth (12.63), there being no significant differences between the independent groups studied ($p = 0.091$).

On the other hand, in the beating phase the preyouth group obtained a higher average range (13.36) than that obtained by the youth group (5.38), there being significant differences ($p = 0.001$) according to the Mann-Whitney statistics; to conclude, in the fall phase there are no significant differences ($p = 0.091$), although the average range of the angle of the spine is greater in the prejuvenile group (11.91) than in the juvenile group (7.38).

When comparing the rest of the variables studied, significant differences were evidenced in the total time of the auction (TTR: $p = 0.033$), the average range being higher in the preyouth group (12.36) than that obtained by the youth group (6.75), while there were no significant differences when comparing the time of preparation to the strike (TPG: $p = 0.717$) and the angle of the body in relation to the center of gravity (ACCG: $p = 0.717$) between independent groups—in both cases there is a lower average range in the preyouth group (TPG = 9.55 and ACCG = 9.59) than in the youth group (TPG = 10.63 and ACCG = 10.56). According to the latest research on biomechanics applied to the auction, we can find that the results allow us to demonstrate what has been exposed by [9] since in his research he concludes that analyzing the technical element allows to know the movement parameters by establishing more specific parameters on the movement patterns present during the execution of the movement. Finally, the values that determine that there are significant differences (beating) correspond to what was stated by Castro et al. [2] since according to his research when comparing the effectiveness of the auction in two different subjects exist differences either by their anthropometric characters physical condition or by the teaching process learning technique.

5. Conclusion

In the angles of the column it was observed that there were no significant differences in the technique of the auction; nevertheless, in the angles of the elbow influence was observed for a better flexion of elbow and placement of the hand, being able to generate a greater impact to the ball and therefore greater difficulties to the opposing defense. It is concluded that since there are significant differences in the striking phase, it is possible to verify the exposition of the studies on biomechanics where the influence of training time, that is, experience, allows movements to be more efficient and generate better results, whether dynamic or kinematic. Say when comparing preyouth players with youth, age, motor development and experience in youth players generate differences, however it should be noted that it is a minimum value which means that during practice and according to the analysis errors can be corrected technicians and in this way obtain a higher level of effectiveness while training the movement patterns that must undergo improvements.

Author details

Erik David García Lasluisa[1*] and Jefferson Michael Vela Rodríguez[2]

1 Upgrade Fitness Center, Quito, Ecuador

2 Unidad Educativa Fiscomisional "Sagrado Corazón de Jesús" Bethlemitas, Tulcán, Ecuador

*Address all correspondence to: erick_c115@hotmail.com

References

[1] Valdes D, Palao J, Bermejo J. Factores Mejorables con el Entrenamiento Asociados a la Efectividad Mecánica del Remate de Voleibol. Entrenamieto Deportivo. 2013;**27**(1)

[2] Garrido Castro J, Gil Cabezas J, da Silva Grigoletto M, Mialdea Baena A, González Navas C. Caracterización cinemática 3D del gesto técnico del remate en jugadoras de voleibol. Revista Andaluza de Medicina del Deporte. 2017

[3] Valladares Iglesias N, Joao P, García Tormo J. Análisis de las variables antropométricas y fisico técnicas en el voleibol femenino. Revista de Ciencias del Deporte. 2016. E-balonmano.com

[4] Pozo Y, Stiven E. Estudios del ataque en el voleibol universitario. E-balonmano.com. Revista de Ciencias del Deporte. 2014;**10**(1):31-39

[5] Esper A. Cantidad y tipos de saltos que realizan las jugadoras de voleibol en el partido. Revista de Educacíon Física y Deportes. 2003;**58**:18-25

[6] Quinita Salas J. Técnica y efectividad biomecánica del remante en el voleibol revisión bibliográfica. Movimiento Humano. 2015;**16**(1):7-12

[7] Hessing W. Voleibol Para Principiantes Entrenamiento técnica y táctica. 4a ed. Barcelona: Paidotribo; 2003

[8] Cardona O, Chalarca Y. Biomecánico de la Ejecución Técnica del Gesto Remate en el Equipo Menores Femenino Perteneciente A la Liga Risaraldense De Voleibol 2012. Universidad Libre Seccional Pereira; 2013

[9] Junior N. Conceptos básicos de la biomecánica del voleibol. Revista Universitaria de la Educacíon Física y el Deporte. 2019;**12**:28-40

[10] Valadés D, Palao J, Femia P, Padial P, Ureña A. Análisis de la técnica básica del remate de voleibol. Rendimiento Deportivo. 2006;**8**

[11] Lucas J. Voleibol iniciacíon y Perfeccionamiento. sexta ed. Barcelona: Paidotribo; 2003

[12] Nápoles Cardona SA, Oris Áreas EY, Medina González F. Aplicación de Ejercicios Para Corregir Errores técnicos en la Fases de la Batida del Remate de Voleibol en Las Jugadoras Auxiliares del Equipo Juvenil Femenino de la Escuela de Iniciación Deportiva (EIDE). EFDeportes; 2013

[13] Cerrato D, Andrés J, Frutos J. Mecánica de la ejecución del remante en veleibol [Mechanics of the spike execution in volleyball]. Movimiento humano. 2013;**5**:33-51

[14] Molina P. Preparación técnica del remate en el Ecuavoley basado en las bases Biomecanicas de su sumilar en le Voleibol [bachelor's thesis]. Universidad Nacional de Chimborazo; 2017

[15] Ramón G, Zapata A. Análisis cinemático de las fases de doble apoyo y de salto, del remante por la zona 4 de los jugadores de la selección masculina de voleibol de antioquia categoría mayores. Educación Física y Deporte. 2006;**25**(1):117-130

[16] Alaiogoikoa Martin, G. (2015). Análisis de la Batida y del Aterrizaje en Jugadoras de Voleibol

[17] Valadés Cerrato D, Palao Andrés J, Fermia Marzo P, Radial Puche P, Ureña Espá A. Validez y fiabilidad del radar para el control de la velocidad del remate en voleibol. Cultura, Ciencia y Deporte. 2007;**2**(6)

[18] Bellendier J. Ataque de rotación en el voleibol un enfoque actualizado.

Lecturas de Educacion Fisica y Deportes. 2002;**51**

[19] Araya C. Análisis biomecánico de la fase del golpe en el remate de voleibol. Lectuas: Educacíon Física y Deportes. 2010;**14**(142)

[20] Suarez GR. Biomecánica Deportiva y Control del Entrenamiento. Medellin: Funámbulos Editores; 2009

[21] Gutiérrez MA. Biomécanica: la física y la fisiología. Madrid: CSIC; 2000

[22] Frutos JB. Revisión del concepto de técnica deportiva desde la perspectiva biomecánica del movimiento. EmásF Revista Digítal de Educación Física. 2013;**25**:15

[23] Jódar XA. Eficiencia y técnica Deportiva Análisis del Movimiento Humano. España: Inde; 1993

[24] Polanco A, Rodriguez C. Modelo Dinámico De Movimientos Humanos. Federacíon Iberoamericana de Ingeniería Mecánica; 2007

[25] Vargas R. Diccionario de Teoria del Entrenamiento Deportivo. Segunda ed. México: UNAM; 2007

[26] Gowitzke BA, Milner M. El cuerpo y sus movimientos bases científicas. Barcelona, España: Paidotrivo; 1999

[27] Kisner C, Colby L. Ejercicio terapéutico Fundametos y técnicas. Barcelona, España: Paidotribo; 2005

[28] Pérez Soriano P, Llana Belloch S. Biomecánica básica aplicada a la activadad fisica y el deporte. Barcelona, España: Paidotribo; 2015

[29] Cromer AH. Física para las ciencias de la vida. In: Barcelona. Reverte: España; 1981

[30] Drauschke K, Kroger C, Schulz A, Utz M. El Entrenador de Voleibol. Barcelona, España: Paidotribo; 2002

Chapter 9

Early Cell Response to Mechanical Stimuli during TBI

Mohammad Mehdi Maneshi and Susan Z. Hua

Abstract

Traumatic brain injury (TBI) refers to brain damage resulting from external mechanical forces such as a blast or crash. The tissue and cell deformations caused by shear forces are the most common pathological features in TBI and lead to long-term symptoms. Our current understanding of TBI derives mainly from *in vivo* studies of poststimulus pathology and the effects on brain function. Little is known about the early responses of brain cells during mechanical stimuli. In this chapter, we evaluate the early cell response to the rapid shear forces *in vitro*. We introduce advanced technologies capable of generating fast shear stimuli mimicking forces occurring in TBI and reporting internal forces in specific proteins at the time of injury. We define the threshold shear forces for calcium influx using an astrocyte model. We describe the spatiotemporal distribution of cytoskeletal forces and correlate them with variations in cell membrane tension. This chapter makes a strong argument that cells' response to external forces is nonlinear. The heterogeneous distribution of cytoskeletal tension and the gradient of protein forces play a key role in the cells' response to mechanical stimuli.

Keywords: traumatic brain injury (TBI), cell mechanics, astrocytes, shear stress, cytoskeletal forces, cell membrane tension

1. Introduction

Traumatic brain injury (TBI) is a form of acquired brain injury that is caused by abrupt external mechanical forces. TBI occurs in both civilians and military veterans from events such as an explosive blast, a blunt impact, and uncontrolled acceleration, deceleration, or rotation of the head. The injury occurs in two phases: a primary injury caused by mechanical forces that occur at the incipient moment of injury and a secondary injury that involves subsequent biochemical and metabolic dysfunction [1–3]. Depending on the type of forces, the injury is normally classified as either a focal injury that results in cerebral contusions in a specific location or a diffuse brain injury that occurs over a widespread area due to shear forces generated by rapid acceleration or deceleration [4–6]. While the focal injuries are common in moderate to severe brain injuries, diffuse injuries are the main cause of mild TBI [7]. The deformations caused by shear forces are more difficult to define because they occur throughout the brain and are often invisible to commonly available imaging techniques at the early stages of injury [8, 9].

Most TBI patients exhibit mild or minimally observable damage upon the initial shock, but serious pathology can develop within hours and days [8, 10, 11]. Cell death may proceed from multiple locations within the brain. Many of these sites contain cells

IntechOpen

with no initially observable injury or proximity to the damage site [12]. Thus, brain cells can respond to low mechanical forces that do not cause an immediate structural damage. Understanding early cell response to mechanical stimuli will help provide valuable insight into the origin and evolution of long-term pathological changes.

Our current understanding of TBI is derived from both *in vitro* and *in vivo* studies. *In vivo* studies provide data on the state of cells well after (hours to days) the injury and probably reflect injuries from downstream processing [13–15]. *In vitro* models, using cultured cells, examine cells' response at the expense of normal physiological interactions [2, 16]. The applied forces for *in vitro* models are often not on the same timescale as TBI. Recent research has begun using the so-called next-generation injury assessment tools to study specific cell responses in real time.

This chapter examines the earliest cell responses during TBI as seen in astrocytes via changes in Ca^{2+} levels *in vitro*. In this chapter, we will introduce advanced technologies that can generate fast shear stimuli mimicking forces that cause TBI and can measure internal forces in specific proteins at the very point of time of injury. We will determine the features of force stimuli that are most susceptible to cells and that lead to long-term alterations. The mechanical properties of cells will be evaluated by real-time measurements of forces in specific cytoskeletal proteins and in cell membranes.

2. Brain structure and traumatic brain injury

2.1 Head structure and brain cells

The brain by its very nature is made of soft substances that are submerged in a thin layer of fluid inside the skull. Underneath the skull, there are multiple layers of tissues separated by intracranial spaces containing water-like liquid, called the cerebrospinal fluid (CSF), as shown in **Figure 1**. It is believed that this fluid plays an important role in the shock-absorbing capacity of the brain. Further inwards, it is the brain itself that consists of a network of functional brain cells. These cells are arranged into several specialized areas, each performing distinct physiological functions.

During TBI, the transient external forces cause linear and angular accelerations, resulting in a range of injuries to the brain. Two main types of injuries may occur: the impact injury due to the brain directly hitting the skull at the point of impact that also generates a whipsaw effect at the opposite side of the brain [17]; and the shear injury due to relative movements between the brain and surrounding tissues and between tissues of different densities [18]. For example, with a sudden acceleration, movement of the brain lags behind that of the skull, producing shear stresses at various interfaces between the brain and the cortical tissues. Similarly, in a deceleration injury, the brain continues its inertial path after the skull has been abruptly halted. Shear injuries commonly occur at gray/white matter junctions, but they are also found in the deeper white matter of the corpus callosum, brain stem, and cerebral cortex [19, 20]. The shear strains and stresses are responsible for producing the loss of consciousness during diffuse axonal injury that accounts for ~60% of hospitalized TBI cases [9, 21].

2.2 Brain cells

Two major cell types in the brain are the neurons and glia. Astrocytes are the most abundant glial cells. While the neurons play the role of processing and transmitting information, astrocytes provide the critical link between the circulatory system and the structural support and the maintenance of neurons (tripartite synapses, neurotransmitter processing, etc.) [22]. **Figure 1** (zoom-in panel) illustrates the configuration of the brain cells. During TBI, astrocytes transmit

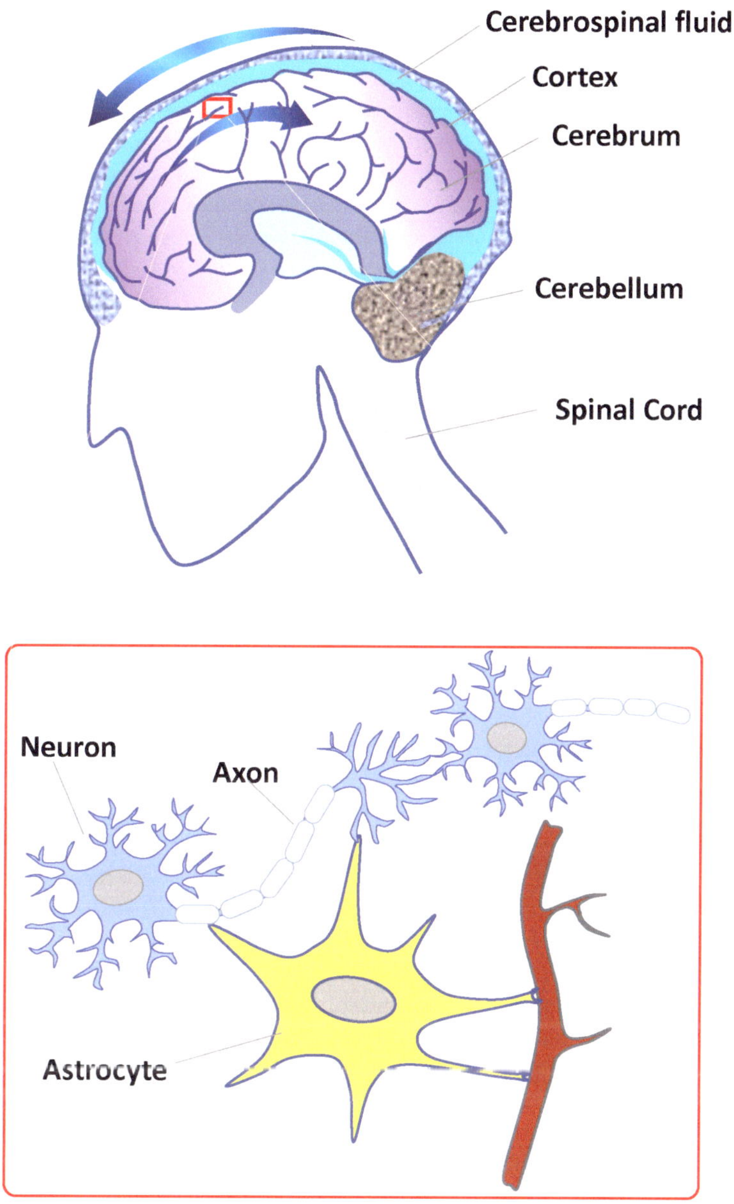

Figure 1.
Schematic illustration of brain structure and brain cells.

mechanical forces to neurons and (by Newton's third law) accept forces from them. They can also propagate damage signals via Ca^{2+} waves, signal to neurons through neurotransmitters, and alter ion concentrations in the extracellular space [22, 23]. By applying a blast-like overpressure to brain slices, it has been shown that astrocyte injury *precedes* the neuronal injury. This raises the possibility that astrocyte Ca^{2+} may be one of the earliest responses modulating the brain function [24].

2.3 Earliest cellular events in injury models

Two early events that are caused by external forces are known: a Ca^{2+} influx and an associated neurotransmitter release [25–27]. An increase in intracellular Ca^{2+} is universally seen in brain cells that have been subjected to mechanical perturbations regardless of the nature of the forces [28–30]. There are many Ca^{2+}-dependent processes that can be activated during TBI, including cell swelling [31–33], cytoskeletal disruption [34–36], membrane poration [37, 38], failure of ATP-dependent membrane pumps [39, 40], and mitochondrial dysfunction and apoptosis [41–44]. The Ca^{2+}-loaded or injured neurons release neurotransmitters including glutamate [45, 46]. The release of glutamate and other excitatory amino acids, in turn, can lead to cytotoxic injury and cell death [47–49]. These occurrences clearly show that the early responses of brain cells to mechanical stimuli are responsible for the subsequent pathology.

3. Shear forces and their assays

3.1 TBI-induced internal forces

The external force results in the development of internal forces within the skull, causing brain injury. Several types of internal forces can occur depending on the type of incidents. For instance, a blast-generated shockwave initially changes the pressure inside the skull. The abrupt motion of the head causes the acceleration or deceleration to generate compression, stretching, and shear forces at multiple places inside the skull [2].

In vitro studies using cultured cells show that excessive pressure (above 10 atm) is required to stimulate a cellular response. This level of pressure is much higher than the overall pressure measured *in vivo* within the brain [50]. Using a blast chamber to apply rapid pressures and shear forces to cell cultures, Raven et al. showed that cells are more responsive to shear forces than to hydrostatic pressures [51].

Tensile forces are primarily localized at impact points and have been commonly studied using stretching methods. In stretching models, cell cultures or tissues are placed on a flexible substrate that can be stretched with a vacuum pulse [52] or by piezoelectric actuators [53]. In these studies, cells produced a measurable Ca^{2+} response only when they experienced very large strains (up to 40%) [54–56]. In contrast, an overall cell deformation of ~4% due to shear pulses was sufficient to cause a similar Ca^{2+} response [57]. These studies show that brain cells are more susceptible to shear forces than other types of internal force.

Most of the studies that have investigated diffusive brain injury due to shear forces have focused on the injury of neurons. A sliding between two tissue layers can break the long thin nerve fibers (called axons) of neurons by their extension across the layers [58, 59]. However, shear stresses have been found to induce drastic responses in cells (neurons and astrocytes) even without any noticeable damage to the axons, leading to cell damage and cell death [60]. In other words, subcellular signaling is likely to be a more causal pathway than obvious physical damage in these cells.

3.2 Microfluidic assays for shear forces

Earlier studies variously used a rotating disk, air blow, or pulsed media to apply shear stresses to the cells [61–63]. Following the application of force stimuli, it was found that transient shear deformations tended to increase the membrane permeability to Ca^{2+} and small dye molecules and decrease the cell viability [64]. A blast chamber was also used to apply rapid pressure and shear forces to cell cultures. It was found that Ca^{2+} response was more sensitive to shear forces than to hydrostatic pressures [51]. These methods provide controlled shear stimuli, followed by live cell measurements. In real scenarios of TBI, the forces increase in milliseconds, and small elastic deformations in cells occur in real time [16]. Thus, the challenge for the shear stress assays is the ability to rapidly ramp the shear forces and an ability to reliably measure cells' response in real time.

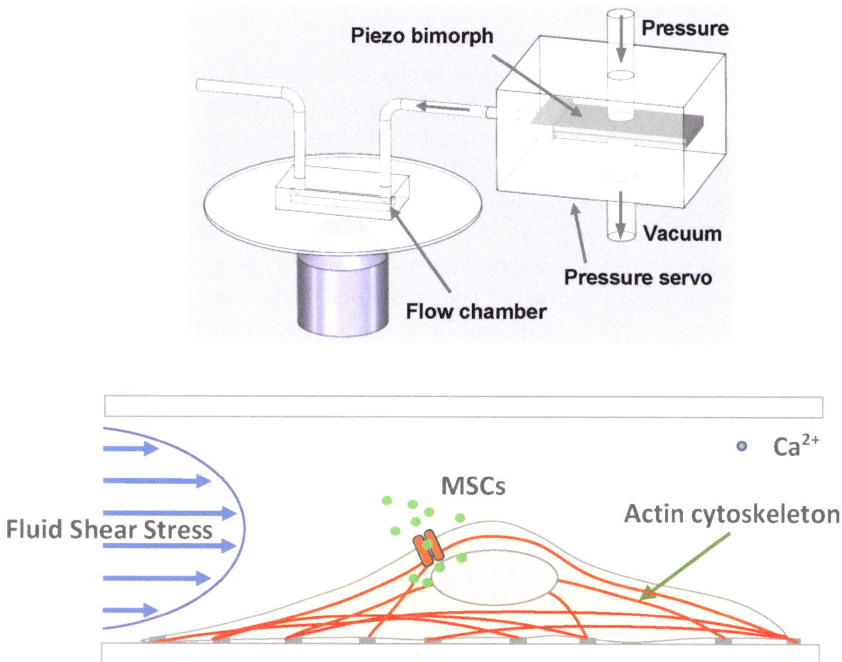

Figure 2.
Schematic of the pressure servo-driven microfluidic chamber. High-speed pressure servo is connected to a microfluidic chamber to generate controlled fluid shear stimuli on cultured cells.

In this chapter, we introduce a high-speed programmable microfluidic assay tool to apply precise shear forces to cultured cells [30]. The fluid shear is controlled by a fast pressure servo [65] that is capable of generating pressure pulses with a time resolution of ~1 ms in the microfluidic chamber [30, 65]. The millisecond resolution allows the ability to program arbitrary waveforms emulating the waveforms recorded in TBI [66]. The microfluidic chamber allows high-resolution optical microscopy, which enables *in situ* measurements of changes in intracellular Ca^{2+} and protein forces in single cells during the very application of forces. **Figure 2** displays the schematic of the experimental setup.

4. The response of cells to transient shear forces

The severity of TBI is conventionally classified as mild, moderate, or severe, depending on the extent of the damage to the brain tissue. In the 1950s, the head injury tolerance to external forces began to be evaluated by dropping embalmed corpse heads on a rigid surface and examining the resulting profile of the skull fracture. Linear skull fracture came to be used as the criterion of injury. These studies led to the first quantitative tolerance curve to be established, now known as the Wayne State Tolerance Curve (WSTC) [67, 68]. This curve describes the influence of magnitude and duration of external loads on the severity of injury to the head.

However, the internal forces are not directly proportional to the external load, which comprise a more complex load profile depending on the type of forces. It is now also known that most cells show a physiological response at much lower internal force levels that are below the injury criteria. Therefore, the thresholds for specific internal forces need to be defined.

4.1 Shear stress thresholds

Using the previously described high-speed servo to generate well-defined shear stresses in a microfluidic chamber, the threshold shear stress to cell response in cultured astrocytes was analyzed. Ca^{2+} rise being the earliest measurable signal in cells, the Ca^{2+} rise in the cells was measured in real time. The thresholds were defined with multiple input parameters including magnitude, duration, and load rise time.

The cells responded reliably to a shear stimulus of 23 dyn/cm^2 with a transient Ca^{2+} increase, but they did not respond up to a pulse of 11.5 dyn/cm^2, thereby setting an effective threshold [30]. However, the amplitude of the force stimuli is not the only parameter, and a variety of relaxation times has been observed. Changes in pulse duration and rise time also affect the subthreshold responses. For example, a low amplitude of short duration (11.5 dyn/cm^2, 10 ms) did not trigger a measurable Ca^{2+} elevation, whereas a pulse width of 1000 ms of the same amplitude did. Together, the two parameters establish a multidimensional threshold matrix for pulses with a square profile, as shown in **Figure 3**. The relative significance of each stimulus

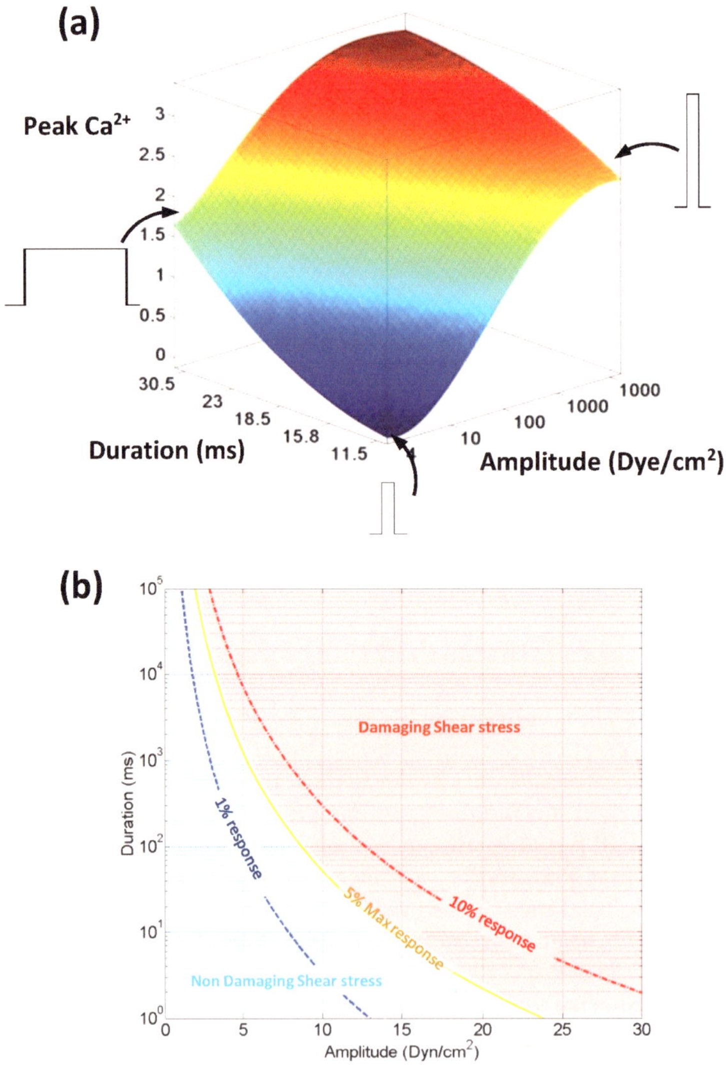

Figure 3.
(a) Profile of shear stimulus thresholds. (b) The damage threshold is defined by a 5% maximum response.

parameter was evaluated by fitting the profile of Ca^{2+} response with a two-state Boltzmann equation, which shows the amplitude is the primary determinant [30].

4.2 Cell response is nonlinear

The cell response to shear stimuli is nonlinear. The nonlinear characteristic of Ca^{2+} response is demonstrated in **Figure 4**. For example, an abrupt increase in shear stress caused the Ca^{2+} rise, but a slow increase to the same amplitude failed to activate the cells, as shown in **Figure 4(a)**. The cell sensitivity decays rapidly with an increase in force rise time. Moreover, a brief mechanical shock is more important than a sustained shear force. As shown in **Figure 4(b)**, at short intervals of 10 ms, the pulse train caused a response that was three times higher than a single pulse of the same amplitude and a total pulse duration of 100 ms [30]. Thus, the total energy input of the stimulus is not the only determinant in Ca^{2+} response; the kinetics of the stimulus (force loading rate) also plays an essential role. This emphasizes the viscoelastic/plastic nature of the force transduction processes in cells [64, 69]. By using sine wave stimuli of different frequencies, it has been shown that cells' response is also frequency dependent; the peaks in response occur at a higher frequency (>80 Hz) or a lower frequency (<20 Hz) [30].

The strain rate dependence of cell injury has been observed in 3D matrix cultured neurons and astrocytes, showing an increase in membrane permeability to small molecules and an increase in post-insult cell death [70, 71]. However, this effect was not observed in hippocampal tissues under biaxial stretch at strain rates ranging from 0.1 to 50 s^{-1} [72]; this rate of ramping is likely outside the sensitivity

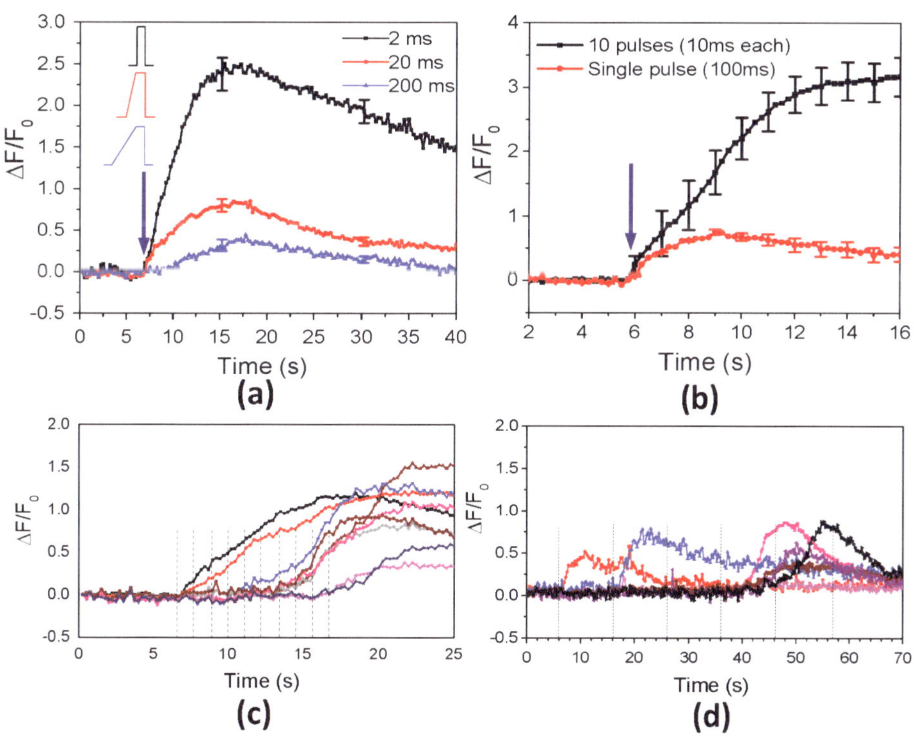

Figure 4.
Nonlinear Ca^{2+} response of cells. (a) Ca^{2+} response to shear pulses of different rise times. (b) Ca^{2+} response to a train of 10 narrow consecutive pulses (10 ms wide) is three times higher than a single broad pulse (100 ms wide) of the same amplitude. (c) and (d) Ca^{2+} response to repeated pulses (11.5 dyn/cm^2, 10 ms) with intervals of 1 s and 10s, respectively, showing that the response is nonlinear.

range of the cells [73]. In addition, different cell types have been found to exhibit different properties, which is not surprising.

4.3 Response to repeated challenges

It has been known for a long time that repeated low-amplitude shocks that never had any clear damage to the cells could be lethal in animal models of TBI [74–76]. This suggests that low subthreshold mechanical forces can be registered and accumulated by the cells. In other words, cells have a long-term pernicious "memory" to repeated stimuli. To study this memory effect, cells were exposed to repetitive low-amplitude stimuli that alone are not able to produce a Ca^{2+} rise. The response to a pulse train with 1 s intervals of 10 ms duration revealed that individual cells were activated at different times. About 20% of the cells responded to the second pulse, while 60% of the cells responded to the sixth and subsequent pulses (**Figure 4(c)**). Reducing the frequency sufficiently can eliminate the response (**Figure 4(d)**). There are two possibilities for this behavior: the plastic mechanical deformation due to cytoskeletal bonds break and reform, or the accumulation of second messenger (Ca^{2+}) pools. Recent studies on cytoskeletal protein forces show that rapid shear stress can generate prolonged cytoskeletal tension, implying that the cytoskeletal deformation plays a key role.

As discussed in detail in the following sections, cells withstand mechanical loading through support from the cytoskeleton consisting of dynamically cross-linked structural proteins [77–80]. The deformation of a cell depends on the intrinsic elastic deformation of fibrous cytoskeletal proteins and plastic deformation involving reversible cross-linking and cytoskeleton reorganization [81, 82]. Slow shear loading engages the plastic processes that significantly modify the local forces around the Ca^{2+} transducers.

The nonlinear response of cells to external forces provides evidence that cells use complex force transduction mechanisms to register the forces. Understanding the force transduction pathways is an important step toward understanding the effectors that lead to TBI pathology.

5. Force transduction mechanisms

While fluid shear stresses act on the apical surface of the cells, this force can be transmitted to force-sensitive molecules that trigger ionic trafficking such as Ca^{2+} influx and other biochemical reactions via a variety of mechanosensitive mechanisms, a process called mechanotransduction. Recent studies suggest that shear stress transduction is mediated by the cytoskeleton, a filamentous network that connects different regions of the cells and holds the mechanical structure of cells [83–85].

5.1 Role of actin cytoskeleton in force transduction

The cytoskeleton is the universal mechanical structure of animal cells and responsible for cell shape, topography, and facilitating cell locomotion [86]. Cells resist the external mechanical forces via the cytoskeleton. The cytoskeleton consists of filamentous proteins that are interlinked to form filaments. Actin filaments are the smallest type of filaments. Single filaments of actin bind together via cross-linking proteins, such as α-actinin, forming a network that connects different regions of the cells. Actin bundles are found across the cell body; actin mesh is also found beneath the cell membrane, supporting the plasma membrane.

Like many other subcellular organelles, the actin cytoskeleton can be affected during TBI. Cytoskeleton damage has been observed in various moderate to severe

TBI conditions [87, 88]. A number of studies now suggest that trauma-induced neuronal cell death may be preceded by the disruption of the neuronal cytoskeleton [89, 90]. Through its connections, the cytoskeleton can also transmit forces to different functional elements of the cell to alter their functions [86, 91–93]. Among them are the mechanosensitive channels (MSCs) that permeate cation ions such as Ca^{2+}.

5.2 Mechanosensitive channels and Ca^{2+} signaling

Mechanosensitive ionic channels are transmembrane proteins that form a pore structure across the cell membrane. These channels are linked to the cytoskeletal proteins; thus, changes in cytoskeletal stresses may open the mechanosensitive channels [94, 95]. Similarly, changes in membrane tension could also alter the channel configuration.

Several MSCs have been identified in astrocytes that are members of the transient receptor potential (TRP) family including TRPV4, TRPC1, TRPC5, and TRPA1 [96–98, 99]. Studies of sensory neurons have suggested that Piezo channels play an important role in brain cells [100, 101]. Piezo-type MSCs are also present in astrocytes that can be inhibited with specific Piezo channel inhibitor [102]. Using an astrocyte model, a recent study has shown that N-Methyl-D-aspartic acid receptors (NMDARs) are the primary Ca^{2+} source in astrocytes and fluid shear stimuli can activate NMDARs in the absence of agonists [57].

The fluid shear forces can alter the channel activities via several mechanisms. They can modify the cytoskeletal stresses, and the cytoskeleton under high tension can pull the channel proteins via their links. Many MSCs are known to link with the actin cytoskeleton with cross-linking proteins [103–105]. In astrocytes and neurons, α-actinin binds to NMDA receptors, providing a mechanical link between NMDA receptors with the underlying cytoskeleton [106]. Shear stress can also cause transient deformation and bending of the lipid bilayer, altering the MSCs directly [107–109]. Most likely, both mechanisms are correlated.

6. Heterogeneous force distribution in single cells subjected to shear pulses

The shear-induced strains and stresses in the cytoskeleton could be the core of the mechanical response of a cell. Since the cytoskeleton is an anisotropic collection of dynamically cross-linked structural proteins, the cytoskeletal stresses are intrinsically nonuniform. This heterogeneity could be responsible for the nonlinear behavior of cells in response to shear stresses.

6.1 Cytoskeletal force measurements

The recent development of force probes based on fluorescence resonance energy transfer (FRET) technique [110–113] provides a powerful tool to observe cytoskeletal protein forces in live cells. The probe consists of two mutant fluorophores, a donor and an acceptor, linked with an elastic structure. Mechanical forces change the distance and/or the angle between the donor and the acceptor, resulting in a change in the energy transfer efficiency from donor to the acceptor [112, 113]. These probes were genetically encoded in cross-linking proteins, such as α-actinin and actin, and they are being used to report the strains in cytoskeletal linking proteins when cells are subjected to shear pulses. Using FRET probes, the distribution of protein forces induced by fluid shear stress has been mapped in real time [114].

6.2 Heterogeneous force distribution in single cells

By inserting the FRET-based force probes in actin-binding protein (α-actinin) in astrocytes, the cytoskeletal protein response to shear stress was measured for the first time. Results show that rapid shear stimuli generate nonuniform protein forces in single cells. **Figure 5** demonstrates that a narrow square shear pulse produced an immediate increase in force in α-actinin at the upstream end of the cell and a small negative force at the downstream end, resulting in a force gradient along the flow direction. In contrast, a slow ramp to the same force amplitude caused a minimal and more uniform change in actinin force. Moreover, the cytoskeleton structure and its prestress states also influence the cell response. The shear pulse produced significant internal force gradients in softer cells that have fewer bundled actin filaments [114]. These experiments demonstrate that uniform shear stress can generate heterogeneous cytoskeletal forces in single cells. Importantly, the time dependence of the stimulus plays a critical role.

While fluid velocity in the flow chamber was precisely controlled, the body stresses produced in the cells depend on cell geometry and the distribution of stresses within the cytoskeleton (we now know that stress distribution is heterogeneous). This heterogeneity likely accounts for the observed variability between cells. Thus, an averaged impact force cannot be used as a parameter to specify the activation of cell activities; the stress at local points can be an order of magnitude higher than the average.

If the local cytoskeletal stresses are the direct cause of Ca^{2+} rise, then they should be correlated in time and space. This correlation cannot be resolved by simple statistical comparisons. The distribution of forces in the structural proteins and the changes in Ca^{2+} was measured simultaneously using a Quad-View optical imaging system. It was found that the activation of Ca^{2+} began in regions of higher strain,

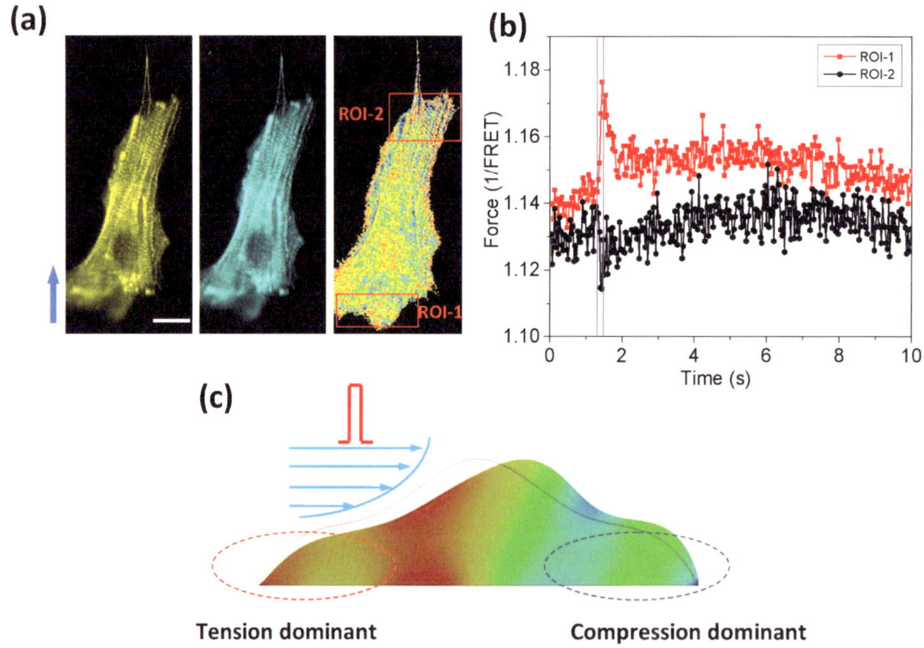

Figure 5.
Rapid fluid shear pulse generates nonuniform force distribution in astrocytes. Panel (a) shows fluorescent images (YFP, CFP) and inverse FRET ratio representing actinin forces in an astrocyte cell. (b) Changes in actinin force in two ROIs in response to a square shear pulse (23 dyn/cm², 15 ms), showing that the narrow shear pulse produce tension in actinin at the upstream edge and compression at downstream edge. (c) Finite element analysis model of a viscoelastic cells under fluid shear stress.

which normally occurred at the upstream edge of the cells and propagated from the edge to the somata of the cell as a wave [114]. In animal models, the NMDAR subunits are found primarily in the astrocytic processes [115]. This finding supports the earlier speculations that applied force may primarily affect the processes where the cytoskeletal structure is most pronounced.

We have noticed that shear stresses above the Ca^{2+} threshold did not disrupt cell adhesions nor produced any visible changes in cytoskeletal structures, indicating that the activation of Ca^{2+}-permeable ion channels require much lower forces than breaking of the bonds or irreversible deformations of the cytoskeleton.

7. Shear-induced cell membrane tension

Since the mechanosensitive ion channels are membrane bound, changes in cell membrane tension inevitably affect the ion channel configurations that mediate Ca^{2+} influx. Indeed, it is known that mechanosensitive channels can be opened by membrane tension [116, 117].

7.1 The cell membrane

The cell membrane consists of a lipid bilayer incorporating the membrane proteins including integral proteins such as transmembrane ion channels and receptors and peripheral proteins that loosely attach to the outer side of the cell membrane. Through the functional proteins, the cell membrane selectively controls the transport of ions, water, and macromolecules between the intracellular and extracellular compartments. Inside the cell, the lipid bilayer intimately adheres to the cortical cytoskeleton that provides the support for membrane topography and integrity [118]. On the outside is a hair-like structure called the glycocalyx. Depending on the cell type and local environments, the cell membrane may have tension at its resting state, called pretension. Several factors contribute to the pretension, including internal forces exerted by the cytoskeleton, osmotic pressure from the cytosol, and the forces resulting from cell-substrate interactions at adhesions that can be passed by cytoskeleton [119].

7.2 Effect of shear stress on cell membrane tension

The bilayer tension can be measured using lipid-soluble molecular rotor probe FCVJ [120] whose mobility is commonly used to extract the lipid bilayer fluidity or viscosity. An increase in bilayer tension increases the fluidity in cell membrane, causing a decrease fluorescent intensity of probes [121]. We measured the bilayer tension in astrocyte membranes using the molecular rotor probe FCVJ with the above-described microfluidic chip. As demonstrated in **Figure 6(a)**, a square pulse of fluid shear (23 dyn/cm^2, 400 ms) generates a gradient in the membrane tension, with higher tension at the upstream edge of the cell and a lower tension (compression) at the distal edge. Both tension and compression recover back to the initial state within ~30 ms. In comparison, the same shear pulse generated a much longer-lasting tension in actinin at the upstream edge of the cell. **Figure 6(b)** and **(c)** illustrates these different characteristics. Interestingly, the membrane tension at the front edge increased much slower than compression at the downstream edge, suggesting that there exists a pretension in the membrane probably via the cortical actin cytoskeleton. The pretension resists the effect of shear force at the upstream edge. In addition, buckling (rapid compression) can occur at the downstream edge. It has been shown that buckling of the lipid membrane can occur at a similar timescale that takes ~150 ms to saturate [122].

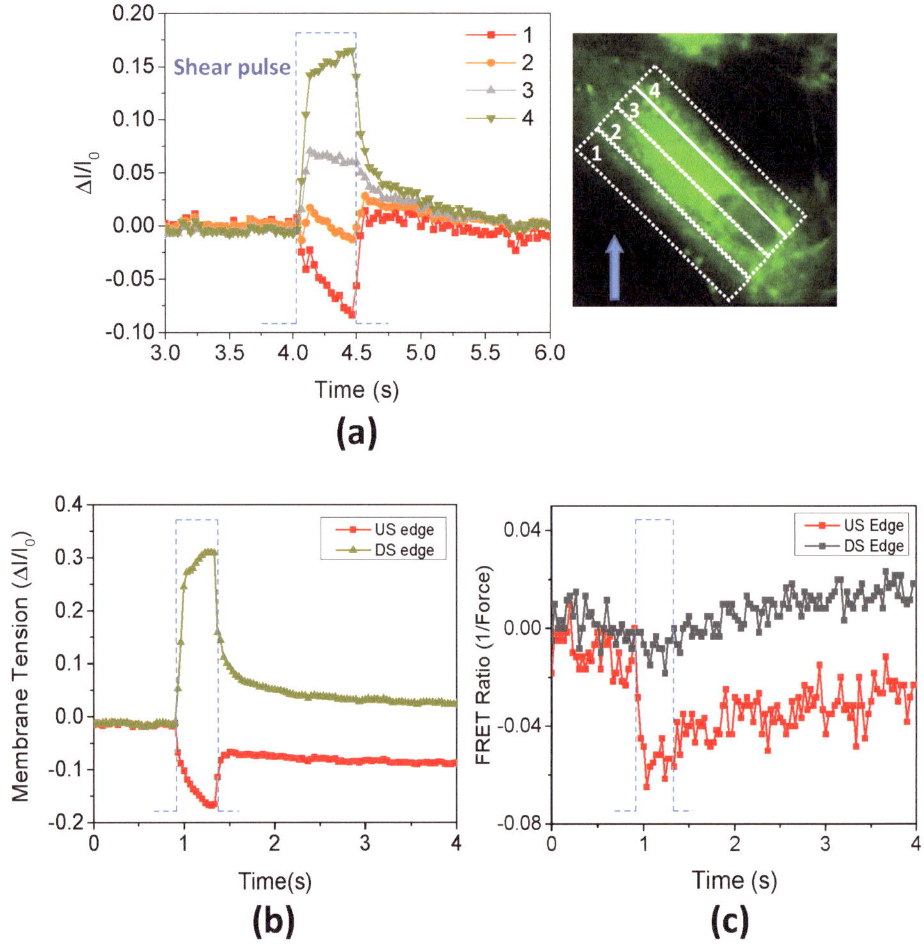

Figure 6.
Change in membrane tension measured using FCVJ molecular rotor incorporated into the astrocyte membrane. (a) Left panel: Membrane tension in response to a shear pulse (23 dyn/cm², 400 ms), where downward inflection indicates tension and upward inflection compression. Labels 1–4 correspond to the regions shown in the image in the right panel. (b) and (c) comparison of membrane tension and actinin force, respectively, in response to a same shear pulse. They show membrane tension at upstream edge increases much slower than actinin force.

A rapid shear pulse generates a gradient in the membrane tension. The spatio-temporal distribution of tension is dependent on the rise time of the shear force. When subjected to a slow ramp-up shear stress, the tension gradient was reduced significantly.

7.3 Membrane tension gradients are coupled to cytoskeletal forces

It has been shown that mechanosensitive channels can be activated by bilayer tension in the lipid vesicles without the cytoskeleton [123]. However, membrane tension measurements show that the membrane tension at the upstream edge increases rather slowly compared with the Ca^{2+} rise. This suggests that the Ca^{2+} channel could not have been activated by the bilayer tension alone and additional mechanisms are likely involved. Slowly ramping the shear stress was not able to change the membrane tension at the upstream edge until it reaches a shear stress threshold. This threshold exactly matches the threshold observed for the rise in the cytoskeletal tension. Therefore, both cytoskeleton and membrane tensions are involved.

Time-dependent analysis shows that shear pulse generates faster and longer-lasting tension in actinin at the front edge of the cell compared to membrane tension, suggesting the cytoskeleton imposes the limiting force at the upstream edge of the cells. In the presence of cytochalasin D that disrupts F-actin, this limit is lifted, and a more predominant membrane tension is seen at the upstream edge compared with controls. While disruption of F-actin diminished tension gradients in the cytoskeleton, it also eliminated the fast influx of Ca^{2+} at the upstream edge of the cell [114]. Thus, cell membrane tension is modulated by the cytoskeleton stresses since the bilayer alone has a minimal ability to bear a large tension gradient.

Moreover, fluid shear by itself is not likely to affect the cell membrane directly since most of the velocity gradient is absorbed by the glycocalyx, leaving little friction at the bilayer [124]. Fluid drag applied to the cell body will pull on the cytoskeleton, thereby increasing its stress [116, 125].

7.4 Effect of shear stress on cell membrane

The effect of shear stress on the lipid membrane response has been variously modeled. The generally proposed model for lipid bilayers is the fluid mosaic model [126]. This model describes the structure of the plasma membrane as a mosaic of components including phospholipids, cholesterol, proteins, and carbohydrates that give the membrane a 2D viscous fluid characteristic. Since the lipid membrane is expected to be intrinsically incompressible, the membrane tension changes could not be explained by solely considering the pure fluid characteristics. Hence, lipid bilayer and MSC interaction via a fluid, or instability of curved stress states of a non-lamellar lipid bilayer, is not easily justifiable within the scope of the fluid mosaic model [127]. Subsequent studies suggest that the lipid membrane is more mosaic rather than fluid [128]. Other models such as the surface model [129] could predict the interaction of lipids and proteins better by allowing both compression and tension changes in the membrane. Based on such models, interactions of lipid and MSCs could be explained by an increase in distance of lipid hydrophobic head groups in the vicinity of proteins and a decrease in outer layer viscosity of membrane [130]. However, without an input from underneath cytoskeleton, the tension gradient in bilayer is minimal.

This leads to the conclusion that shear stress generates tension gradients in the cell cortex and that membrane tension gradients are coupled to cytoskeletal forces to mediate Ca^{2+} influx. The time-dependent membrane tension gradient is modulated by the intact cytoskeleton forces.

8. Evaluation of astrocytic Ca^{2+} response to shear in brain slices

In the brain, neurons and astrocytes are intimately connected and function through a three-dimensional circuit that passes information waves. The interplay between them is evident in bidirectional glutamatergic astrocyte-neuron signaling in a Ca^{2+}-dependent fashion. A common consequence of TBI is the alternations of this information flow.

While *in vitro* experiments described above permit high-resolution measurements, the environment differs from *in vivo*. To better approximate the *in situ* conditions, and to see how much of the *in vitro* results are applicable *in vivo*, similar sets of experiments can be performed to brain slides, since slides would contain the native cell types and their local environments as *in vivo*. To a closer approximation, a modified shear chamber was used to apply fluid shear stress to mechanically stimulate the slices.

The Ca^{2+} response in acute slices from rats is demonstrated in **Figure** 7, which shows how shear stimuli modulate Ca^{2+} response in cells under physiological

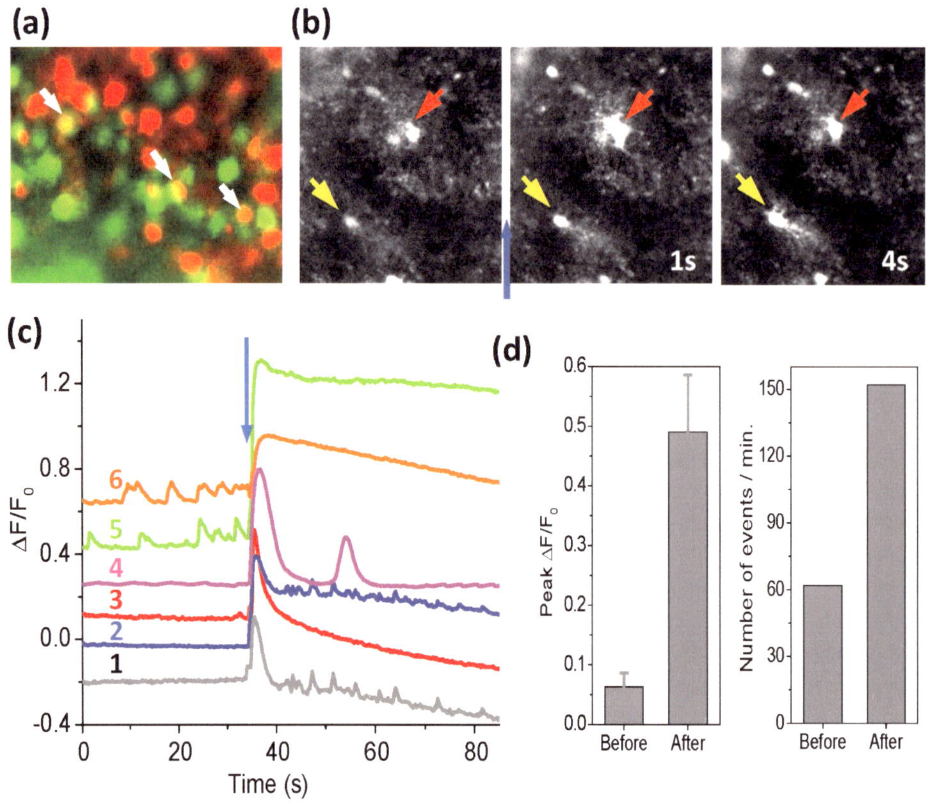

Figure 7.
Astrocyte Ca^{2+} response to a shear pulse in a hippocampal slice. (a) The slice is co-loaded with Flou-4 (green) and SR101 (red). (b) Time sequence of Ca^{2+} images showing Ca^{2+} peaks at different times in selected cells. (c) Typical traces of astrocyte Ca^{2+} response of individual cells. (d) Statistics of peak amplitudes and frequency.

conditions (with the caveat that these may be treated as samples of extreme TBI). To discriminate the astrocytes from neurons, the slices were loaded with SR101 that serves as a marker for astrocytes, as shown in **Figure 7(a)**. Shear stimulated slices showed an acute Ca^{2+} increase in selected cells that peaked in 1 to 4 s and returned to baseline levels within 20 s, consistent with observations in cell cultures. Most of the cells showed one dominant peak, but some (~20% of cells) responded with multiple peaks (trace 4, **Figure 7(c)**). The average peak Ca^{2+} was much higher than the spontaneous Ca^{2+} transients that were 10–20% of the shear-induced peak (traces 5 and 6, **Figure 7(c)**). These Ca^{2+} peaks were eliminated with 10 μM Gd^{3+}, which is a nonspecific MSC blocker. This confirms the observations that shear stress-induced transient Ca^{2+} peaks are via MSCs.

It is worthwhile pointing out that the fluid shear stress generates well-controlled forces on the apical surface of cell cultures. In the slide experiments, once the deformation reaches a deeper layer of the cells, the poroelastic nature of the tissue will also modify the forces, so the stimulus profile itself may change as it propagates.

9. Conclusion

Using advanced technologies that can generate fast shear stimuli mimicking forces that cause TBI, we have demonstrated that cell response to mechanical stimuli is nonlinear and the features of the stimuli play a critical role. Using FRET-based

DOI: http://dx.doi.org/10.5772/intechopen.93295

force probes, the nonlinearity is shown to be a direct result of nonuniform force distribution within the cytoskeleton. Rapid shear pulses generate a heterogeneous distribution of cytoskeletal forces in cells, in both time and space. The cytoskeletal forces and their modulation on cell membrane tension open MSCs that mediate Ca^{2+} response. These early response signals can be small and transient. However, the integration of these signals leads to pathology and the progression of TBI.

Acknowledgements

This work was supported by the National Institutes of Health grant NS085517 and National Science Foundation grants CMMI-1537239 and CMMI-2015964.

Author details

Mohammad Mehdi Maneshi and Susan Z. Hua*
Department of Mechanical and Aerospace Engineering, Department of Physiology and Biophysics, University at Buffalo, Buffalo, New York, USA

*Address all correspondence to: zhua@buffalo.edu

References

[1] North SH, Shriver-Lake LC, Taitt CR, Ligler FS. Rapid analytical methods for on-site triage for traumatic brain injury. Annual Review of Analytical Chemistry (Palo Alto, California). 2012;**5**:35-56

[2] Chen YC, Smith DH, Meaney DF. In-vitro approaches for studying blast-induced traumatic brain injury. Journal of Neurotrauma. 2009;**26**(6):861-876

[3] Meaney DF, Smith DH. Biomechanics of concussion. Clinics in Sports Medicine. 2011;**30**(1):19-31

[4] Garman RH, Jenkins LW, Switzer RC 3rd, Bauman RA, Tong LC, Swauger PV, et al. Blast exposure in rats with body shielding is characterized primarily by diffuse axonal injury. Journal of Neurotrauma. 2011;**28**(6):947-959

[5] DeKosky ST, Blennow K, Ikonomovic MD, Gandy S. Acute and chronic traumatic encephalopathies: Pathogenesis and biomarkers. Nature Reviews. Neurology. 2013;**9**(4):192-200

[6] Sanborn B, Nie X, Chen W, Weerasooriya T. Inertia effects on characterization of dynamic response of brain tissue. Journal of Biomechanics. 2012;**45**(3):434-439

[7] Graham DI, Adams JH, Nicoll JA, Maxwell WL, Gennarelli TA. The nature, distribution and causes of traumatic brain injury. Brain Pathology. 1995;**5**(4):397-406

[8] Kan EM, Ling EA, Lu J. Microenvironment changes in mild traumatic brain injury. Brain Research Bulletin. 2012;**87**(4-5):359-372

[9] Johnson VE, Stewart W, Smith DH. Axonal pathology in traumatic brain injury. Experimental Neurology. 2013;**246**:35-43

[10] Koliatsos VE, Cernak I, Xu L, Song Y, Savonenko A, Crain BJ, et al. A mouse model of blast injury to brain: Initial pathological, neuropathological, and behavioral characterization. Journal of Neuropathology and Experimental Neurology. 2011;**70**(5):399-416

[11] Saljo A, Svensson B, Mayorga M, Hamberger A, Bolouri H. Low-level blasts raise intracranial pressure and impair cognitive function in rats. Journal of Neurotrauma. 2009;**26**(8):1345-1352

[12] Ellis EF, McKinney JS, Willoughby KA, Liang S, Povlishock JT. A new model for rapid stretch-induced injury of cells in culture: Characterization of the model using astrocytes. Journal of Neurotrauma. 1995;**12**(3):325-339

[13] Kamnaksh A, Kovesdi E, Kwon SK, Wingo D, Ahmed F, Grunberg NE, et al. Factors affecting blast traumatic brain injury. Journal of Neurotrauma. 2011;**28**(10):2145-2153

[14] Bramlett HM, Dietrich WD. Progressive damage after brain and spinal cord injury: Pathomechanisms and treatment strategies. Progress in Brain Research. 2007;**161**:125-141

[15] Rubovitch V, Ten-Bosch M, Zohar O, Harrison CR, Tempel-Brami C, Stein E, et al. A mouse model of blast-induced mild traumatic brain injury. Experimental Neurology. 2011;**232**(2):280-289

[16] Morrison B 3rd, Elkin BS, Dolle JP, Yarmush ML. *In vitro* models of traumatic brain injury. Annual Review of Biomedical Engineering. 2011;**13**:91-126

[17] Ommaya AK, Hirsch AE. Tolerances for cerebral concussion from head impact and whiplash in primates. Journal of Biomechanics. 1971;**4**(1):13-21

[18] Parker RS. Concussive Brain Trauma: Neurobehavioral Impairment & Maladaptation. Boca Raton, FL, USA: CRC Press; 2011

[19] Ivancevic VG. New mechanics of traumatic brain injury. Cognitive Neurodynamics. 2009;**3**(3):281-293

[20] Taber KH, Warden DL, Hurley RA. Blast-related traumatic brain injury: What is known? The Journal of Neuropsychiatry and Clinical Neurosciences. 2006;**18**(2):141-145

[21] van Eijck MM, Schoonman GG, van der Naalt J, de Vries J, Roks G. Diffuse axonal injury after traumatic brain injury is a prognostic factor for functional outcome: A systematic review and meta-analysis. Brain Injury. 2018;**32**(4):395-402

[22] Chen Y, Swanson RA. Astrocytes and brain injury. Journal of Cerebral Blood Flow and Metabolism. 2003;**23**(2):137-149

[23] Niggel J, Sigurdson W, Sachs F. Mechanically induced calcium movements in astrocytes, bovine aortic endothelial cells and C6 glioma cells. The Journal of Membrane Biology. 2000;**174**(2):121-134

[24] Sarntinoranont M, Lee SJ, Hong Y, King MA, Subhash G, Kwon J, et al. High-strain-rate brain injury model using submerged acute rat brain tissue slices. Journal of Neurotrauma. 2012;**29**(2):418-429

[25] Fineman I, Hovda DA, Smith M, Yoshino A, Becker DP. Concussive brain injury is associated with a prolonged accumulation of calcium: A 45Ca autoradiographic study. Brain Research. 1993;**624**(1-2):94-102

[26] Shapira Y, Yadid G, Cotev S, Shohami E. Accumulation of calcium in the brain following head trauma. Neurological Research. 1989;**11**(3):169-172

[27] Mills LR, Velumian AA, Agrawal SK, Theriault E, Fehlings MG. Confocal imaging of changes in glial calcium dynamics and homeostasis after mechanical injury in rat spinal cord white matter. NeuroImage. 2004;**21**(3):1069-1082

[28] LaPlaca MC, Thibault LE. Dynamic mechanical deformation of neurons triggers an acute calcium response and cell injury involving the N-methyl-D-aspartate glutamate receptor. Journal of Neuroscience Research. 1998;**52**(2):220-229

[29] Weber JT, Rzigalinski BA, Ellis EF. Traumatic injury of cortical neurons causes changes in intracellular calcium stores and capacitative calcium influx. The Journal of Biological Chemistry. 2001;**276**(3):1800-1807

[30] Maneshi MM, Sachs F, Hua SZ. A threshold shear force for calcium influx in an astrocyte model of traumatic brain injury. Journal of Neurotrauma. 2015;**32**(13):1020-1029

[31] Bullock R, Maxwell WL, Graham DI, Teasdale GM, Adams JH. Glial swelling following human cerebral contusion: An ultrastructural study. Journal of Neurology, Neurosurgery, and Psychiatry. 1991;**54**(5):427-434

[32] Kimelberg HK, Rutledge E, Goderie S, Charniga C. Astrocytic swelling due to hypotonic or high K+ medium causes inhibition of glutamate and aspartate uptake and increases their release. Journal of Cerebral Blood Flow and Metabolism. 1995;**15**(3):409-416

[33] Povlishock JT, Becker DP, Cheng CL, Vaughan GW. Axonal change in minor head injury. Journal of Neuropathology and Experimental Neurology. 1983;**42**(3):225-242

[34] Monnerie H, Tang-Schomer MD, Iwata A, Smith DH, Kim HA, Le Roux PD. Dendritic alterations

after dynamic axonal stretch injury *in vitro*. Experimental Neurology. 2010;**224**(2):415-423

[35] Tang-Schomer MD, Patel AR, Baas PW, Smith DH. Mechanical breaking of microtubules in axons during dynamic stretch injury underlies delayed elasticity, microtubule disassembly, and axon degeneration. The FASEB Journal. 2010;**24**(5):1401-1410

[36] Siedler DG, Chuah MI, Kirkcaldie MT, Vickers JC, King AE. Diffuse axonal injury in brain trauma: Insights from alterations in neurofilaments. Frontiers in Cellular Neuroscience. 2014;**8**:429

[37] Farkas O, Lifshitz J, Povlishock JT. Mechanoporation induced by diffuse traumatic brain injury: An irreversible or reversible response to injury? The Journal of Neuroscience. 2006;**26**(12):3130-3140

[38] Lafrenaye AD, McGinn MJ, Povlishock JT. Increased intracranial pressure after diffuse traumatic brain injury exacerbates neuronal somatic membrane poration but not axonal injury: Evidence for primary intracranial pressure-induced neuronal perturbation. Journal of Cerebral Blood Flow and Metabolism. 2012;**32**(10):1919-1932

[39] Unterberg AW, Stover J, Kress B, Kiening KL. Edema and brain trauma. Neuroscience. 2004;**129**(4):1021-1029

[40] Davalos D, Grutzendler J, Yang G, Kim JV, Zuo Y, Jung S, et al. ATP mediates rapid microglial response to local brain injury *in vivo*. Nature Neuroscience. 2005;**8**(6):752-758

[41] Pivovarova NB, Andrews SB. Calcium-dependent mitochondrial function and dysfunction in neurons. The FEBS Journal. 2010;**277**(18):3622-3636

[42] Orrenius S, Zhivotovsky B, Nicotera P. Regulation of cell death: The calcium-apoptosis link. Nature Reviews. 2003;**4**(7):552-565

[43] Singh IN, Sullivan PG, Deng Y, Mbye LH, Hall ED. Time course of post-traumatic mitochondrial oxidative damage and dysfunction in a mouse model of focal traumatic brain injury: Implications for neuroprotective therapy. Journal of Cerebral Blood Flow and Metabolism. 2006;**26**(11):1407-1418

[44] Sauerbeck A, Hunter R, Bing G, Sullivan PG. Traumatic brain injury and trichloroethylene exposure interact and produce functional, histological, and mitochondrial deficits. Experimental Neurology. 2012;**234**(1):85-94

[45] Yi JH, Hazell AS. Excitotoxic mechanisms and the role of astrocytic glutamate transporters in traumatic brain injury. Neurochemistry International. 2006;**48**(5):394-403

[46] Persson L, Hillered L. Chemical monitoring of neurosurgical intensive care patients using intracerebral microdialysis. Journal of Neurosurgery. 1992;**76**(1):72-80

[47] Lee JM, Zipfel GJ, Choi DW. The changing landscape of ischaemic brain injury mechanisms. Nature. 1999;**399**(6738 Suppl):A7-A14

[48] Ankarcrona M, Dypbukt JM, Bonfoco E, Zhivotovsky B, Orrenius S, Lipton SA, et al. Glutamate-induced neuronal death: A succession of necrosis or apoptosis depending on mitochondrial function. Neuron. 1995;**15**(4):961-973

[49] McAllister TW. Neurobiological consequences of traumatic brain injury. Dialogues in Clinical Neuroscience. 2011;**13**(3):287-300

[50] Murphy EJ, Horrocks LA. A model for compression trauma:

Pressure-induced injury in cell cultures. Journal of Neurotrauma. 1993;**10**(4):431-444

[51] Ravin R, Blank PS, Steinkamp A, Rappaport SM, Ravin N, Bezrukov L, et al. Shear forces during blast, not abrupt changes in pressure alone, generate calcium activity in human brain cells. PLoS One. 2012;**7**(6):e39421

[52] Morrison B III, Meaney DF, McIntosh TK. Mechanical characterization of an *in vitro* device designed to quantitatively injure living brain tissue. Annals of Biomedical Engineering. 1998;**26**(3):381-390

[53] King AI, Ruan J, Zhou C, Hardy WN, Khalil T. Recent advances in biomechanics of brain injury research: A review. Journal of Neurotrauma. 1995;**12**(4):651-658

[54] Rzigalinski BA, Weber JT, Willoughby KA, Ellis EF. Intracellular free calcium dynamics in stretch-injured astrocytes. Journal of Neurochemistry. 1998;**70**(6):2377-2385

[55] Alford PW, Dabiri BE, Goss JA, Hemphill MA, Brigham MD, Parker KK. Blast induced phenotypic switching in cerebral vasospasm. Proceedings of the National Academy of Sciences of the United States of America. 2011;**108**(31):12705-12710

[56] Singh P, Doshi S, Spaethling JM, Hockenberry AJ, Patel TP, Geddes-Klein DM, et al. N-methyl-D-aspartate receptor mechanosensitivity is governed by C terminus of NR2B subunit. The Journal of Biological Chemistry. 2012;**287**(6):4348-4359

[57] Maneshi MM, Maki B, Gnanasambandam R, Belin S, Popescu GK, Sachs F, et al. Mechanical stress activates NMDA receptors in the absence of agonists. Scientific Reports. 2017;7:39610

[58] Pena A, Pickard J, Stiller D, Harris N, Schuhmann M. Brain tissue biomechanics in cortical contusion injury: A finite element analysis. In: Intracranial Pressure and Brain Monitoring XII. Vienna, Austria: Springer; 2005. pp. 333-336

[59] Zhang L, Yang KH, King AI. A proposed injury threshold for mild traumatic brain injury. Journal of Biomechanical Engineering. 2004;**126**(2):226-236

[60] Ravin R, Blank PS, Steinkamp A, Rappaport SM, Ravin N, Bezrukov L, et al. Shear forces during blast, not abrupt changes in pressure alone, generate calcium activity in human brain cells. PLoS One. 2012;**7**(6):e39421

[61] LaPlaca MC, Thibault LE. Dynamic mechanical deformation of neurons triggers an acute calcium response and cell injury involving the N-methyl-D-aspartate glutamate receptor. Journal of Neuroscience Research. 1998;**52**(2):220-229

[62] Chung RS, Staal JA, McCormack GH, Dickson TC, Cozens MA, Chuckowree JA, et al. Mild axonal stretch injury *in vitro* induces a progressive series of neurofilament alterations ultimately leading to delayed axotomy. Journal of Neurotrauma. 2005;**22**(10):1081-1091

[63] Chung RS, Staal JA, McCormack GH, Dickson TC, Cozens MA, Chuckowree JA, et al. Mild axonal stretch injury *in vitro* induces a progressive series of neurofilament alterations ultimately leading to delayed axotomy. Journal of Neurotrauma. 2005;**22**(10):1081-1091

[64] Cullen DK, Vernekar VN, LaPlaca MC. Trauma-induced plasmalemma disruptions in three-dimensional neural cultures are dependent on strain modality and rate. Journal of Neurotrauma. 2011;**28**(11):2219-2233

[65] Besch SR, Suchyna T, Sachs F. High-speed pressure clamp. Pflügers Archiv / European Journal of Physiology. 2002;**445**(1):161-166

[66] Taylor PA, Ludwigsen JS, Ford CC. Investigation of blast-induced traumatic brain injury. Brain Injury. 2014;**28**(7):879-895

[67] Gurdjian ES, Webster JE. Surgical considerations in cerebro-vascular accidents. Transactions of the American Neurological Association. 1950;**51**:121-123. Discussion 3-5

[68] Lissner HR, Lebow M, Evans FG. Experimental studies on the relation between acceleration and intracranial pressure changes in man. Surgery, Gynecology & Obstetrics. 1960;**111**:329-338

[69] LaPlaca MC, Cullen DK, McLoughlin JJ, Cargill RS 2nd. High rate shear strain of three-dimensional neural cell cultures: A new *in vitro* traumatic brain injury model. Journal of Biomechanics. 2005;**38**(5):1093-1105

[70] Cullen DK, Vernekar VN, LaPlaca MC. Trauma-induced plasmalemma disruptions in three-dimensional neural cultures are dependent on strain modality and rate. Journal of Neurotrauma. 2011;**28**(11):2219-2233

[71] Tamayol A, Bahrami M, editors. Analytical solutions for laminar fully-developed flow in microchannels with non-circular cross-section. In: ASME 2009 Fluids Engineering Division Summer Meeting. New York, USA: American Society of Mechanical Engineers; 2009

[72] Cater HL, Sundstrom LE, Morrison B. Temporal development of hippocampal cell death is dependent on tissue strain but not strain rate. Journal of Biomechanics. 2006;**39**(15):2810-2818

[73] Yeung EW, Whitehead NP, Suchyna TM, Gottlieb PA, Sachs F, Allen DG. Effects of stretch-activated channel blockers on [Ca2+] i and muscle damage in the mdx mouse. The Journal of Physiology. 2005;**562**(2):367-380

[74] Koch W, Filene W. Beiträge zur experimentellen Chirurgie. Über die commotio cerebri. Arch Klin Chir. 1874;**17**:P190

[75] Shultz SR, Bao F, Omana V, Chiu C, Brown A, Cain DP. Repeated mild lateral fluid percussion brain injury in the rat causes cumulative long-term behavioral impairments, neuroinflammation, and cortical loss in an animal model of repeated concussion. Journal of Neurotrauma. 2012;**29**(2):281-294

[76] Smith DH, Johnson VE, Stewart W. Chronic neuropathologies of single and repetitive TBI: Substrates of dementia? Nature Reviews. Neurology. 2013;**9**(4):211-221

[77] Verma D, Ye N, Meng F, Sachs F, Rahimzadeh J, Hua SZ. Interplay between cytoskeletal stresses and cell adaptation under chronic flow. PLoS One. 2012;**7**(9):e44167

[78] Meng F, Sachs F. Orientation-based FRET sensor for real-time imaging of cellular forces. Journal of Cell Science. 2012;**125**(3):743-750

[79] Meng F, Sachs F. Measuring strain of structural proteins *in vivo* in real time. In: Kohl P, Sachs F, Franz MR, editors. Cardiac Mechano-Electric Coupling and Arrhythmia: From Pipette to Patient. Oxford, UK: Oxford University Press; 2011. pp. 431-434

[80] Rahimzadeh J, Meng F, Sachs F, Wang J, Verma D, Hua SZ. Real-time observation of flow-induced cytoskeletal stress in living cells. American Journal of Physiology. Cell Physiology. 2011;**301**(3):C646-CC52

[81] Gardel M, Shin J, MacKintosh F, Mahadevan L, Matsudaira P, Weitz D. Elastic behavior of cross-linked and bundled actin networks. Science. 2004;**304**(5675):1301-1305

[82] Bursac P, Lenormand G, Fabry B, Oliver M, Weitz DA, Viasnoff V, et al. Cytoskeletal remodelling and slow dynamics in the living cell. Nature Materials. 2005;**4**(7):557-561

[83] Ito S, Majumdar A, Kume H, Shimokata K, Naruse K, Lutchen KR, et al. Viscoelastic and dynamic nonlinear properties of airway smooth muscle tissue: Roles of mechanical force and the cytoskeleton. American Journal of Physiology. Lung Cellular and Molecular Physiology. 2006;**290**(6):L1227-L1237

[84] Alenghat FJ, Nauli SM, Kolb R, Zhou J, Ingber DE. Global cytoskeletal control of mechanotransduction in kidney epithelial cells. Experimental Cell Research. 2004;**301**(1):23-30

[85] Vogel V, Sheetz M. Local force and geometry sensing regulate cell functions. Nature Reviews. Molecular Cell Biology. 2006;**7**:265-275

[86] Stricker J, Falzone T, Gardel ML. Mechanics of the F-actin cytoskeleton. Journal of Biomechanics. 2010;**43**(1):9-14

[87] Deng Y, Thompson BM, Gao X, Hall ED. Temporal relationship of peroxynitrite-induced oxidative damage, calpain-mediated cytoskeletal degradation and neurodegeneration after traumatic brain injury. Experimental Neurology. 2007;**205**(1):154-165

[88] Graham DI, Saatman KE, McIntosh TK. The neuronal cytoskeleton is at risk after mild and moderate brain injury. Journal of Neurotrauma. 1998;**15**(12):1047-1058

[89] Newcomb JK, Kampfl A, Posmantur RM, Zhao X, Pike BR, Liu S-J, et al. Immunohistochemical study of calpain-mediated breakdown products to α-spectrin following controlled cortical impact injury in the rat. Journal of Neurotrauma. 1997;**14**(6):369-383

[90] Posmantur R, Newcomb J, Kampfl A, Hayes R. Light and confocal microscopic studies of evolutionary changes in neurofilament proteins following cortical impact injury in the rat. Experimental Neurology. 2000;**161**(1):15-26

[91] Orr AW, Helmke BP, Blackman BR, Schwartz MA. Mechanisms of Mechanotransduction. Developmental Cell. 2006;**10**(1):11-20

[92] Gardel ML, Shin JH, MacKintosh FC, Mahadevan L, Matsudaira P, Weitz DA. Elastic behavior of cross-linked and bundled actin networks. Science. 2004;**304**(5675):1301-1305

[93] Gardel ML, Nakamura F, Hartwig JH, Crocker JC, Stossel TP, Weitz DA. Prestressed F-actin networks cross-linked by hinged filamins replicate mechanical properties of cells. Proceedings of the National Academy of Sciences of the United States of America. 2006;**103**(6):1762-1767

[94] Mills J, Mandel L. Cytoskeletal regulation of membrane transport events. The FASEB Journal. 1994;**8**(14):1161-1165

[95] Hayakawa K, Tatsumi H, Sokabe M. Actin stress fibers transmit and focus force to activate mechanosensitive channels. Journal of Cell Science. 2008;**121**:496-503

[96] Ostrow LW, Sachs F. Mechanosensation and endothelin in astrocytes--hypothetical roles in CNS pathophysiology. Brain Research. Brain Research Reviews. 2005;**48**(3):488-508

[97] Suchyna TM, Tape SE, Koeppe RE, Andersen OS, Sachs F, Gottlieb PA. Bilayer-dependent inhibition of mechanosensitive channels by neuroactive peptide enantiomers. Nature. 2004;**430**(6996):235-240

[98] Ostrow LW, Suchyna TM, Sachs F. Stretch induced endothelin-1 secretion by adult rat astrocytes involves calcium influx via stretch-activated ion channels (SACs). Biochemical and Biophysical Research Communications. 2011;**410**(1):81-86

[99] Shigetomi E, Tong X, Kwan KY, Corey DP, Khakh BS. TRPA1 channels regulate astrocyte resting calcium and inhibitory synapse efficacy through GAT-3. Nature Neuroscience. 2012;**15**(1):70-80

[100] Coste B, Mathur J, Schmidt M, Earley TJ, Ranade S, Petrus MJ, et al. Piezo1 and Piezo2 are essential components of distinct mechanically activated cation channels. Science. 2010;**330**(6000):55-60

[101] Blumenthal NR, Hermanson O, Heimrich B, Shastri VP. Stochastic nanoroughness modulates neuron–astrocyte interactions and function via mechanosensing cation channels. Proceedings of the National Academy of Sciences. 2014;**111**(45):16124-16129

[102] Suchyna TM, Tape SE, Koeppe RE 2nd, Andersen OS, Sachs F, Gottlieb PA. Bilayer-dependent inhibition of mechanosensitive channels by neuroactive peptide enantiomers. Nature. 2004;**430**(6996):235-240

[103] Müller P, Langenbach A, Kaminski A, Rychly J. Modulating the actin cytoskeleton affects mechanically induced signal transduction and differentiation in mesenchymal stem cells. PLoS One. 2013;**8**(7):e71283

[104] Wechsler A, Teichberg VI. Brain spectrin binding to the NMDA receptor is regulated by phosphorylation, calcium and calmodulin. The EMBO Journal. 1998;**17**(14):3931-3939

[105] Martinac B. The ion channels to cytoskeleton connection as potential mechanism of mechanosensitivity. Biochimica et Biophysica Acta (BBA) - Biomembranes. 2014;**1838**(2):682-691

[106] Sekino Y, Kojima N, Shirao T. Role of actin cytoskeleton in dendritic spine morphogenesis. Neurochemistry International. 2007;**51**(2-4):92-104

[107] Haidekker MA, L'Heureux N, Frangos JA. Fluid shear stress increases membrane fluidity in endothelial cells: A study with DCVJ fluorescence. American Journal of Physiology. Heart and Circulatory Physiology. 2000;**278**(4):H1401-H14H6

[108] White CR, Frangos JA. The shear stress of it all: The cell membrane and mechanochemical transduction. Philosophical Transactions of the Royal Society, B: Biological Sciences. 2007;**362**(1484):1459-1467

[109] Bloom M, Evans E, Mouritsen OG. Physical properties of the fluid lipid-bilayer component of cell membranes: A perspective. Quarterly Reviews of Biophysics. 1991;**24**(03):293-397

[110] Meng F, Suchyna TM, Sachs F. A fluorescence energy transfer-based mechanical stress sensor for specific proteins *in situ*. The FEBS Journal. 2008;**275**(12):3072-3087

[111] Grashoff C, Hoffman BD, Brenner MD, Zhou R, Parsons M, Yang MT, et al. Measuring mechanical tension across vinculin reveals regulation of focal adhesion dynamics. Nature. 2010;**466**(7303):263-266

[112] Meng F, Sachs F. Orientation-based FRET sensor for real-time imaging of cellular forces. Journal of Cell Science. 2012;**125**(Pt 3):743-750

[113] Guo J, Wang Y, Sachs F, Meng F. Actin stress in cell reprogramming. Proceedings of the National Academy of Sciences of the United States of America. 2014;**111**(49):E5252-E5261

[114] Maneshi MM, Sachs F, Hua SZ. Heterogeneous cytoskeletal force distribution delineates the onset Ca(2+) influx under fluid shear stress in astrocytes. Frontiers in Cellular Neuroscience. 2018;**12**:69

[115] Conti F, DeBiasi S, Minelli A, Melone M. Expression of NR1 and NR2A/B subunits of the NMDA receptor in cortical astrocytes. Glia. 1996;**17**(3):254-258

[116] Cox CD, Bae C, Ziegler L, Hartley S, Nikolova-Krstevski V, Rohde PR, et al. Removal of the mechanoprotective influence of the cytoskeleton reveals PIEZO1 is gated by bilayer tension. Nature Communications. 2016;7:10366

[117] Sachs F. Mechanical transduction by ion channels; a cautionary tale. World Journal of Neurology. 2015;5(3):74-87

[118] Kapus A, Janmey P. Plasma membrane--cortical cytoskeleton interactions: A cell biology approach with biophysical considerations. Comprehensive Physiology. 2013;**3**(3):1231-1281

[119] Kozlov MM, Chernomordik LV. Membrane tension and membrane fusion. Current Opinion in Structural Biology. 2015;**33**:61-67

[120] Haidekker MA, Ling T, Anglo M, Stevens HY, Frangos JA, Theodorakis EA. New fluorescent probes for the measurement of cell membrane viscosity. Chemistry & Biology. 2001;**8**(2):123-131

[121] Markin VS, Sachs F. Free volume in membranes: Viscosity or tension? Open Journal of Biophysics. 2015;**05**(03):80-83

[122] Helfer E, Harlepp S, Bourdieu L, Robert J, MacKintosh F, Chatenay D. Buckling of actin-coated membranes under application of a local force. Physical Review Letters. 2001;**87**(8):088103

[123] Kloda A, Lua L, Hall R, Adams DJ, Martinac B. Liposome reconstitution and modulation of recombinant N-methyl-D-aspartate receptor channels by membrane stretch. Proceedings of the National Academy of Sciences. 2007;**104**(5):1540-1545

[124] Florian JA, Kosky JR, Ainslie K, Pang Z, Dull RO, Tarbell JM. Heparan sulfate proteoglycan is a mechanosensor on endothelial cells. Circulation Research. 2003;**93**(10):e136-ee42

[125] Akinlaja J, Sachs F. The breakdown of cell membranes by electrical and mechanical stress. Biophysical Journal. 1998;**75**(1):247-254

[126] Singer S, Nicolson GL. The fluid mosaic model of the structure of cell membranes. Science. 1972;**175**(4023):720-731

[127] Bagatolli LA, Mouritsen OG. Is the fluid mosaic (and the accompanying raft hypothesis) a suitable model to describe fundamental features of biological membranes? What may be missing? Frontiers in Plant Science. 2013;**4**:457

[128] Engelman DM. Membranes are more mosaic than fluid. Nature. 2005;**438**(7068):578-580

[129] Brown MF. Curvature forces in membrane lipid–protein interactions. Biochemistry. 2012;**51**(49):9782-9795

[130] Haswell ES, Phillips R, Rees DC. Mechanosensitive channels: What can they do and how do they do it? Structure. 2011;**19**(10):1356-1369